TEN REASONS FOR NOT NAMING YOUR CAT CALCULUS

TEN REASONS FOR NOT NAMING YOUR CAT CALCULUS

AND OTHER FASCINATING MATH ADVENTURES

JERRY FARLOW

THALES HOUSE PRESS

Dedication

To Susan, who had the poor judgement to marry someone like me, but I'm awfully glad she did.

Contents

Preface

I was slightly baffled a while back when a student came rushing in my office, but instead of trying to entice me to cough up the blueprints of an upcoming exam or reciting a well-prepared oration for turning in a late homework, she asked a rather curious question.

"Do you remember an old student of yours from 25 years ago?" she said.

Normally, I don't remember the names of students I had the previous semester, but for some reason I remembered this one student.

"Why yes, I remember her," I admitted.

"*She's my mom!*" she blurted out.

And to make matters worse, the incident didn't take place last week. *It was twenty years ago!* So to make sure that another student doesn't come rushing in my office and ask me if I remember so-and-so, then blurt out, *"She's my grandma!"* I decided to take the safe route, retire, and, as they say, get the hell outta Dodge.

A short while after retirement I began to see the benefits of retirement: snoozing in my favorite rocker, lots of relaxation, lots of contemplation, and even more *boredom*. Observing that I was slowly going stark-raving mad, my wife suggests that after teaching and doing mathematical research for 50 years, I might write a book of some kind. Just to verify that I had been doing *something* in the past half-century, she made the wry suggestion.

Little did she know that for the past fifty years, I *had* been writing a book, I just didn't know it. Often, after a long week slaving over a hot differential equations class, I would crash out on Friday night and spew out a chunk of words about some mundane, albeit mathematical, topic.

Although fifty years will result in a pile of words, getting the average book editor, no doubt an English Lit major from a humanities college, to get exercised over such a collection of stories, categorized as "math miscellanea" is a task not taken lightly. Once the average editor sees a book so classified, a rejection letter is not far behind. However, there are a few enlightened editors who see the value of presenting mathematics in various shades of grey, so here we are.

Ten Reasons for Not Naming Your Cat Calculus consists of short mathematical adventures ranging from satire to serious mathematics to downright silliness.

The story *Caution: Burned-Out Hypotheses Ahead*, turns the tables on my old automobile mechanic, Rodney Camrod, whose inspiration came while sitting in the waiting room of a car garage, waiting for Rodney to install a new drive-shaft, knowing full well he had a good-sized shaft for me as well.

Of course I had to include, *Baseball's Magic Number: Smaller is Better*, which describes how baseball's magic number is computed, a skill every red-blooded American should know.

Feel free to skip over any story that doesn't fit your fancy or contains too much or too little mathematical minutia. Perhaps, however, there are stories which resonate with your curiosity and you will gain something from the experience. Or if you have nothing better to do, feel free to read the entire book, cover to cover. Enjoy.

Jerry Farlow
Professor Emeritus of Mathematics
University of Maine

ΦΣΞΛΘΩ

Forward

The happiest day of my life was the day when this little *tome* was finally shipped off to the printers! Maybe some grass will finally get mowed around this place, or maybe someone, whose name I will not divulge, will take out the garbage!

If I have to answer one more question about whether the period goes before or after the quotation mark, I'll go stark raving mad. Or the difference between "affect" and "effect," sheesh, don't math professors ever learn any grammar?

Now that he's done with his *magnum opus*, as he calls it, he'll no doubt migrate back to his usual headquarters in front of the TV, watching weekend football, and demanding the chips and dips keep coming.

My only desire is that the dear reader of his *tour de force*, gets as much enjoyment from it as I do, knowing it's 100% done.

Susan Farlow
Author's wife

ΦΣΞΛΘΩ

About the Author

My publisher told me that this was the place in the book where I should include all the pretentious crap about myself that I could muster. He said just don't end a sentence with a preposition (or proposition) or mix up commonly misused words like "elicit" and "illicit," but other than that to illicit anything I could think of.

My writing career began on a dark and stormy night when my famed travelwriter wife suggested our funds were trending low, that I might do my part by writing a best-selling [cough] math book.

I said that might be a good idea since my experience in the writing field was established long ago when I spilled a bottle of writing ink on the dress of my first-grade teacher, Miss. Altman, an innocent accident for which she had no quarter. After that came college and my English professor, Mr. Kerrigan, who gave me a D in Eng Comp 101 for my refusal to follow all those fuddy-duddy old rules about composition and English usage.

But things turned around for me after I became a college professor and began accumulating desk drawers of pedagogical tailings. In my attempt to pass on learned words of wisdom to future generations, I spent weekends rummaging through pages of old notes and lesson plans, summarizing their contents in 1,000-word essays. I was ill at ease over the less-than-Harvard-level of scholarship of my writings, and so I published them anonymously in various less-than-Harvard-level publications under the name *Nats Wolraf*, the mirror image of my first and last names, *Stan Farlow*. Although Nat's career as a purveyor of mathematical nuances never reached the stratosphere like those of fellow Mainer,

Stephen King, they did provide motivation for Nats to carry on. Once I got infected with that time-honored habit, I reached out to the textbook field, where no doubt, at this very moment there are legions of students poking about in one of my seminal texts on calculus, finite math and partial differential equations, no doubt using my name in vain. Did they actually think the answers to the problems came with the book?

But for those readers who are not into the more serious recesses of the Queen of the Sciences, this book might just tickle their fancy, and do a minuscule amount of educating in the process. And if not, there are others that might fit the bill, including the following [cough] bestsellers.

- The Girl Who Ate Equations for Breakfast, Thales House Press

- Mathematics:Ain't There an App for That? Thales House Press

- Partial Differential Equations, Dover Publications

- Paradoxes in Mathematics, Dover Publications

Jerry Farlow

ΦΣΞΛΘΩ

TEN REASONS FOR NOT NAMING YOUR CAT CALCULUS

I thank Peter Roget for letting me express my angst when I say the worst, lousy, rotten, dire, god-awful mistake I ever made was when I took my cat Calculus to college with me. I had Calculus from when I was a young boy and I couldn't bear to part with him. Here are just a few prob-

---

lems I experienced during my college years with Calculus.

One weekend I called my girlfriend and asked her if she wanted to come over. She said yes. I told her we could spend the afternoon with Calculus. She never came and I never saw her again.

The next day, I took Calculus to the vet for some shots. On the way home, I met a friend who asked me how my classes were going. I told him everything was going great, my English professor was helping me with my writing, I had a tutor for my Spanish class, and I had just been to the vet who helped me with Calculus.

A couple of weeks later, my drinking buddies called and asked if I wanted to go to a party that weekend. They said there would be a lot of cute girls there. I told them fine and added that I had spent the day playing with Calculus. That was the last time they invited me.

In my off-time, I volunteered in a nursing home, assisting the elderly residents. I suggested to the manager that Calculus might be a fun diversion for the elderly, that even a 90-year old could spend quality time with Calculus. She told me to continue with the bedpans.

Calculus studying calculus

At the end of a semester, I had a nervous break-down and went to the college psychologist. He worried about me getting over-stressed and a lit-tle wacky and asked if I had a pet that might take my mind off my studies. I told him I had a cat and my neighbor had a greyhound who loved Calcu-lus. The doctor doubled my prescription.

Later that semester, I dated a cheerleader who asked me if I was well rounded and not just a nerdy math type. I told her not to worry that I was well-rounded and liked many things. I told her I spent a great deal of my time with Calculus since I was one year old. I never saw her after that.

The doctor sent me to the campus infirmary for psychological evaluation, wanting to learn more about my early life. I told them I had a normal childhood even from the age of two, spending many happy hours with in my baby crib with Calculus. They gave me some more pills.

My place? We can play with Calculus for hours!

My advisor asked me at the end of my freshman year how I planned on spending the summer. I told him I wanted to write a children's book about Calculus.

By the time I was a senior, I gained somewhat of a reputation as a math whiz and tutor. One day I was contacted by a freshman who was worried about upcoming exams. I told her not to worry, that if she came to my apartment I would help her. She said she didn't have much money for tutoring, but said she really wanted to know calculus. I told her I wouldn't charge for the tutoring, but if she was serious about getting to know Calculus, it would be a good idea for her to bring along a tin of sardines and a bag of cat liter.

At the end of my college career I asked my adviser to write a letter of recommendation for me for grad school. He asked me what calculus had taught me in my four years of college. I told him Calculus taught me that I'd never get a dog. I never got into grad school.

After graduation from college I had an interview with a company. The interviewer asked how I might use mathematics in solving the company's problems. I told him I'd love to apply mathematics in solving their problems, and in fact Calculus might come in handy in rodent control in the company's warehouse. I never got the job.

$$\Pi\psi\Delta\Phi\theta\Lambda$$

2

DUGAN O'KEEFE: DEFENDER OF MATHEMATICAL LAWS

The *story you are about to read is true, only the names have not been changed to protect the innocent. The fat, ugly woman in the second chapter is based on my aunt Emma; the boisterous, overbearing man that appears sporadically is a mirror image of my uncle Harold, and the acid-tongued,*

7

nagging Amazon is, amazingly, almost a spitting image of my mother-in-law.

———¤¤¤⌶⌶⌶¤¤¤———

The name's O'Keefe, Dugan O'Keefe. I carry a badge. I'm the fuzz, a gumshoe, the man. I work the night division out of the Twenty-third Precinct Math Patrol (dummm-da-dum-dum). It was a Wednesday night 2:55 A.M., I was on stakeout in a real tough part of town — keeping an eye out for a no-good, two-bit vermin called Johnny Weasel, alias the "Weasel."

I've been shacked up with the front seat of my `76 Caddie for six hours straight, chewin' on a piece of cold Domino's pizza washed down with some warm Pabst beer. The upholstery of my car is stained more than the bedding in a puppy mill. I pop open another Pabst, take a drag from a stale Marlboro, and blow a ring of smoke through my nose. A few minutes later I crack open the last of the suds and flip the empty out the window. The Weasel is starting to get on my nerves.

Holy Toledo! How do people manage to read this stuff? I've written only two paragraphs, and I'm already going into Mickey Spillane shock. But hey, if you're willing to read it, I'll write it. Anyways, I gotta story to tell.

Suddenly, I spot the maggot slithering down the street with bottle of gin under one arm and a math book under the other. I'm going to nail the cockroach. He's broken every math law in the book, even the granddaddy of them all — the Commutative Law of Multiplication. *God, what's the world coming to?* I follow him to a run-down flop house and slam him against the wall.

"Awwwww Jeez, O'Keefe," he moaned. "I shoulda known it was you."

"Shut up!" I tell him, putting on my gutsiest snarl. "You know the drill, spread `em." I shake him down. As I suspected, the usual tools of the trade: two Slim Jims, a gooseneck, and a dog-eared copy of Farlow's, *Mathematics for Beginners.*

"Ah ha," I said, thumbing through the battered

pages of the book. "Just what do we have here?" I opened the book to the chapter, *Basic Laws of Arithmetic*. Just as I suspected, he crossed off each one of them and replaced them with phonies.

"Violating some of the laws of arithmetic again, eh Weasel?" I snarled.

"Hey, I'm clean I tell ya! I'm clean!"

"Don't give me that, you know what I'm talking about."

Just as I thought, the Weasel had marked off the Commutative Law of Multiplication

$$ab = ba$$

and scribbled in the phony

$$ab \neq ba$$

"Are ya thinking of violating the Commutative Law of Multiplication, Weasel?" I shoved the page in his face. "And I suppose you're not the sack of slime that's been scribbling $ab \neq ba$ in the washrooms over at the bus station, are ya?"

"I've never been near the place," the Weasel groaned, feeling his arm being twisted off.

"And you're not the sewage-scum that's been hanging around the schoolyard over on 43rd street, telling the little kiddies $ab \neq ba$, are ya?

"You can't pin that on me," the Weasel squealed like the rat he was.

"I don't want you putting any weirdo ideas in those kidde's heads, ya gittin my drift? Those little kiddies should know they can multiply numbers any order they want, it don't make no difference in the outcome."

"But not all things are the same," the Weasel argued, "but it's not always true that doing things in opposite order gives the same thing."

"Ya just can't quit can ya weasel," I said, "just give me one."

"Goin outside and emptying the garbage, that

ain't the same as emptying the garbage and goin outside. So hah, *ab* is not always equal to *ba*," the Weasel said smugly.

"I should take ya down town for that," I warned, "we're not talkin about stuff like that, we're talkin math things, like numbers. The Commutative Law of Multiplication says it don't matter what order you multiply numbers, like 3 x 2 and 2 x 3 are the same."

"But it's not even true for all numbers," the Weasel insisted. I heard about this dame, a Bugs Maxwell, over on 63rd Street that knows about numbers where multiplication ain't commutative," the Weasel said.

"Nice try, Weasel," I said. "Now why don't you and I take a ride downtown."

"No, no, it's true! That Maxwell dame has got these numbers."

"Ok," I gave in, "You've earned yourself some time, Weasel. You'd just better pray to your Rat of Jehovah this Maxwell dame comes up with the goodies."

———¤¤¤ΞΞΞ¤¤¤———

A few hours later I find myself hammering on the door of a Bugs Maxwell. It was a long shot, but I'm a fair cop so I was giving the Weasel his last chance to keep out of the slammer. If this broad really had discovered some non-commutative numbers, then I owed him that much.

The door opened and before me stood a petite, no slender, no willowy brunette. She had puppy dog eyes the color of unset opals and was wearing a long black negligee that did things to me. She gave me a sultry look as if she knew me from somewhere and said icily, "*Let me guess, a copper?*"

"I could go for you, doll," I said in a professional manner.

"Why don't ya go oil your gun or something," she said as she slammed the door in my face.

———

"Depends if ya got any three-in-one oil, sugar," I said kicking the door open.

"Now you be a good little girl and we'll get along just fine."

"I like a man that knows what he wants," her attitude changed on a dime. She flung herself around my neck and pulled herself tight against me. "Take me sweetie and do some bad things to me," she purred. I tried to get away from her with some quick moves, but she mistook my movements for passionate sparring, and she jumped on me, pushing me onto the sofa.

"I want to ask you something about algebra," I said.

"Just as I thought, a math cop," she barked.

"Enough to know the difference between statistics and *sadistics*," I confessed.

"Exactly what is the difference?" she asked.

"Well, uh, ... it has, uh, well,... ," I quavered.

"Can we get back to the Weasel?"

"So you're looking for some numbers that satisfy

$$ab \neq ba$$

she said.

"That's it," I replied.

"I have some numbers that *act* like numbers, but they are not numbers that you are familiar," she said.

"What do you mean?" I asked.

"Well, like ordinary numbers, you can carry out all usual operations like addition and multiplication," she said.

"If they quack like numbers, they're numbers," I said.

"But since they are not exactly like ordinary numbers you are familiar, we must make up new definitions of addition and multiplication," she said.

So what do these new numbers look like," I asked.

She said she started with four *basic* "numbers," she denoted by

$$1, \; i, \; j, \; k$$

The number 1 is the usual 1 from grade-school arithmetic, but the i, j, and k are new numbers she made up. She said she could call them anything she wanted, but $i, j,$ and k were convenient symbols. She then defined the product of any two of these numbers as the intersection of the two respective numbers in the table below.

For example,

$$i^2 = j^2 = k^2 = -1$$

and

$$i \times j = k, \; j \times i = -k$$

and so on.

X	1	i	j	k
1	1	i	j	k
i	i	-1	k	$-j$
j	j	$-k$	-1	i
k	k	j	$-i$	-1

Quaternion multiplication
table

"So those are your new numbers?" I asked. "They don't look like numbers to me. What are i, j and k anyway?"

"*They are my numbers!*" she said. "Numbers are anything you want them to be, they don't have to be 1, 2, 3, ..."

"Are there only four numbers

$$1, i, j, k$$

in your arithmetic?" I asked sheepishly.

"No," she said, but she added that all the other numbers can be written in terms of them. All the new numbers are of the form

$$a + bi + cj + dk$$

where a, b, c and d are any real numbers you wish. In other words, there are an infinite number of these new numbers. Two typical "numbers" in her new system would be

$$3 + 3i + 4j + 6k$$
$$6 - 2j + 3k$$

You can make your own up if you wish. What's more, addition of these numbers is defined in a natural way. For example, the sum of these two new numbers would be

$$(3 + 3i + 4j) + (-2j + 3k) = 3 + 3i + 2j + 3k$$

Other numbers of this type are added in the same way. Note that it doesn't matter the order you add these numbers. In other words, addition is commutative for these new numbers.

She also told me how she could use her multiplication table to find the product of any two of her numbers. But in this new arithmetic system of "generalized numbers," the operation of multiplication does not satisfy the Commutative Law of Multiplication, $ab = ba$, although they do satisfy other laws of arithmetic, like

- Commutative Law of Addition: $a + b = b + a$

- Distributive Law: $a(b + c) = ab + ac$

"Well, I'll be," I said dumbfounded. "When did you come up with those strange numbers?"

"I didn't," she said. "They've been around for 150 years. They're called quaternions. Haven't you ever heard of the Irish mathematician, William Rowan Hamilton, who discovered them in 1843? He thought of them while crossing a bridge in Dublin. Ain't you Irish?"

"Uh," I stammered.

The interesting thing about quaternions is that until Hamiltonian discovered them, everyone thought multiplication must be commutative in all legitimate arithmetic systems. Hamilton's discovery showed that there exist legitimate arithmetic systems that behave differently from grade-school arithmetic. These systems opened people's eyes to new novel algebraic systems, which are important not only in pure mathematics but in many practical areas including computer science and engineering.

"Quaternions huh?" I confessed. "I'd never heard of them before."

"I like a man that can admit his mistakes, "she leaped on me again and pushed me onto the sofa.

Just another night on the Math Patrol I was thinking.

ΠψΔΦθΛ

3

CAUTION: BURNED-OUT HYPOTHESES AHEAD

The inspiration for this story came while sitting in the waiting room of a car garage, waiting for the mechanic to install a new drive-shaft in my car, suspecting he had a good-sized shaft for me as well.

Good things just don't fall into your lap in my world so when my automobile mechanic Rodney Camrod came running hell-bent into my office with a faulty *Mean-Value Theorem*, my eyes perked right up.

"Hmmmmmmmmm," I said looking out from under a 23rd -edition of Thomas' *Calculus and Analytic Geometry*. "Now that could be a problem."

"I'm not sure," Rodney admitted, "but there does seem something wrong with uh, what's it called, the Mean-Value Theorem?"

"Well then," I said, rolling up my sleeves and spreading out a dozen pages of his proof on my desk. "Let's take a peek in there." It didn't take me very long to find the problem.

"Hmmmmmmm," I said.

"What is it?" Rodney asked.

"Well friend you've got a peck of trouble. What you have is a faulty contra-positive. You should be arrested walking around with a proof like that. Who put that proof in there, an engineer?"

"A faulty counter-positive?" he quavered.

Sorry, but I'm goin' have to replace that busted Mean Value Theorem.

"The word is *contra-positive*," I smiled to myself.

"I don't know anything about mathematical proofs," Camrod admitted. I smiled again.

"Well, that's what I'm here for, I'll have it fixed in no time. I'll just yank out that old contra-positive and replace it with a brand new direct proof."

"A direct proof," Camrod asked. "Why is that?"

Most people look at a faulty contra-positive and say why, I look at a faulty contra-positive and say $1489, plus tax.

"Can't you get it working for under $150," Rodney begged. "I'm really short on cash."

"'Fraid not." I had to give him the honest truth, "I'll have to replace all the nasty quantifiers. That's a big job."

"But I don't have $1489," Camrod said, but I really need this proof."

"Well," I said in a deliberate manner, "I guess I could install a proof by *reductio ad absurdum* for about $650, but we mathematicians frown on them."

"I think I'll get another estimate if you don't mind," Camrod said snidely.

"Be my guest," I said as Camrod slunk out the door.

The next day he was back.

"I guess I'll take that *reductio* thing," Rodney said, "but $650 is as far as I can go," he groaned.

With that, I began the reconstruction of his

proof. The next day Camrod was back, looking over my shoulder.

"How's it going in there?" he asked.

"I'm afraid we've got more problems," I said, giving him the bad news.

"I thought you said all it needed was a new proof!"

"You can't always tell unless you really get inside and start looking around," I said. "After closer examination, you find all sorts of things, messy notation, oversimplified conclusions, and in your case, burned-out hypotheses."

"*Burned-out hypotheses?*" Rodney gasped.

"That's right," I said, "It takes more than simple continuity to prove the Mean-Value Theorem. This isn't freshman calculus class, you know." Rodney's face turned red.

"I'm afraid it's going to run you about $3500 to get this little baby in shape," I said. "And why we're at it, I think we'd better tear out those existential quantifiers and put in some shiny new universal ones."

"Why, I just had those quantifiers checked last month," Rodney whined, getting a little testy.

"Look buddy, are you questioning my integrity?" I sneered, taking a step forward.

"No, no, I believe you," he said, backing away.

"I got my reputation to uphold here," I told him.

"Ok, ok, I'll take the $3500 overhaul," he gave in. But I have to go to the bank and borrow the money."

"Never mind," I advised, "I got my own little loan company right here, just for your convenience. I call it *Exponential Financing*. I'm sure you'll find it of great interest."

$$\Pi\psi\Delta\Phi\Theta\Lambda$$

4

THE SKELETON IN GOD'S CLOSET

———

"Yes?" the chief ombudsman said, looking up from his desk.

"May I help you?"

"Oh," the intruder paused at the door.

"Don't be afraid, I'm here to help."

"I've never come here before. I've never had anything to say, but now it can't wait," the man at the door said nervously.

"Fiddlesticks, just sit yourself down and tell me your problem." With that, the large pleasant-looking man behind the desk rose and nodded in the direction of a nearby chair. The man sat down and began to speak.

"I don't think you realize sir, but we have a problem."

"What kind of problem? By the way, what should I call you?"

"Willoby, sir, 67701."

"Well, Mr. Willoby, what is our problem?" the ombudsman, said as he pulled Willoby's dossier from a filing cabinet."

It's Hell."

"Hell?"

"Yes sir, Hell," Willoby paused, then stuttered

nervously, "uh ... it's ... it's ... getting better than ... uh .... Heaven."

"What?" the ombudsman said, "How can anything be better than here?"

"Well," Willoby said uneasily not looking up., "It all started with the mathematicians. Hell is full of them. There are topologists, algebraists, and even some, uh, applied mathematicians. Why, there are so many applied mathematicians and engineers down there they've installed air-conditioning. It's a pleasant 70 degrees. They say it's more comfortable there than here."

"*What?*" the ombudsman repeated, not believing his ears.

"It's true, it's true, and that's just the beginning, they,... "

"Wait Willoby, let me get this straight, you say that there are so many mathematicians in Hell they've got air-conditioning?"

"That's right," Willoby said, "I have this friend ..."

"All right, I believe you, you wouldn't be here if you were a liar."

Willoby smiled, he had led an exemplary life on earth and now in Heaven he was a model angel, but he felt it was his duty to inform God's chief ombudsman and troubleshooter about these latest goings-on at the other end of the Universe. There had been a lot of scuttlebutt around Heaven, but the rumors hadn't reached the top.

"That's right," Willoby continued, "but it's a lot more serious than the air-conditioning. Everyone has the right to make themselves comfortable. I understand it was awfully hot down there."

"I suppose," said the ombudsman.

"But, over the last few years other kinds of earthlings, besides these mathematicians, have gone to Hell after their life on earth."

"What kinds?" asked the ombudsman.

"Like I said, it all started with the mathematicians but later there were writers, painters, scientists, you name 'em, all kinds of people. In fact," Willoby cleared his throat, "even some, some, uh, theologians."

"Theologians?"

"It's true, it's true, and I can prove it."

"Ok, ok," the ombudsman interjected. "I believe you. But how could a theologian not get into Heaven?"

"That's just the point, sir," Willoby said. He stammered for a moment.

"Well?" the ombudsman waited anxiously.

"Well, you see sir, uh, they don't all want into Heaven anymore." Again, the ombudsman stared at Willoby in disbelief.

"What? Why would anyone go to Hell before Heaven? It's against the laws of the universe," the ombudsman said.

"They say Hell is more interesting than Heaven," Willoby continued. "The mathematicians hold weekly seminars and lectures down there. It's very exciting."

The ombudsman gasped.

"In fact," Willoby continued, starting to gain confidence, "during the past twenty-five years, architects, urban-planners, biologists, computer scientists you name 'em, they're all opting for Hell. They've designed complete cities using the latest technology. They have internet and can send e-mail messages from one end of Hell to the other. They've even installed a new computerized network and have their own website

http://www.come-on-down.hell

Computerized trains zip people from one end of Hell to the other. Botanists have even developed new strains of plants that grow in that awful soil. They even have Facebook! They e-mail kitten videos from one end of Hell to the other. They've turned it into a paradise, it's a,a,uh, a heaven."

"*Heaven!?*" the ombudsman shouted.

"Is that what they call it, a *Heaven*!?"

"Well, uh" Willoby tried to defend himself, then continued, "we're not getting the right kind of people anymore.

"You're saying we should change our admittance requirements?" the ombudsman asked.

"It wouldn't work, "Willoby said, "there are so many advertising executives down there that the instant an earthling dies, he's bombarded with a media blitz."

"Nooooooo!?" the ombudsman gasped.

"It's true, at the exact instant of death, a person gets an e-mail from Hell with a message like. *You're in good hands with Lucifer,* or *When you say Beelzebub you've said it all,* or even *With a name like Mephistopheles it has to be good.*"

The ombudsman was stunned. Something clearly had to be done to stop this brain drain to Hell.

After a few moments of silence, Willoby spoke again.

"There is still another aspect which I haven't mentioned, far more serious."

The ombudsman's face grew pale.

"More serious?" he gulped.

"Yes," Willoby said. "It concerns some angels here in Heaven."

"Go on."

Hmmmmmm. what's going on down there?

"Well," Willoby said, not looking directly into

the ombudsman's eyes, "some, uh, some want to be transferred."

"Transferred!?" the ombudsman shouted, jumping out of his chair. "Transferred? Is that what they call it!?"

"Some do," Willoby stuttered.

"That does it!" the ombudsman shouted. "This conversation is going straight to the Big Guy himself. We're going to nip this problem in the bud!" With that, he grabbed the red phone and dialed #1.

"Doris, is the Boss in? This is urgent," the ombudsman said, holding the receiver tight against his ear.

After a moment, he spoke.

"I see, oh really, that's strange. Well, let me know when he returns." With that, he swung around and looked intently at Willoby.

"What is it?" Willoby asked.

"Strange," the ombudsman said, "God's not in his office this afternoon and said he wouldn't be back until Monday." The ombudsman looked puzzled.

"What is it?" Willoby pressed.

"Well," the ombudsman said. "When he left the office, he said something strange."

"What?" Willoby asked.

"He just said

$$e^{i\pi} + 1 = 0$$

and walked out the door. *"I wonder what he meant by that?"*

ΠψΔΦθΛ

5

THWARTED BY LAPLACE'S EQUATION

You might say Mel was as happy as a mosquito in a nudist colony. Yes siree. It had been an extraordinary run and we're not talking daily double, grand quinella or super perfecta. We're talking about *Irish sweepstakes.*

Melvin had gone nine for nine and so with the poker hot he decided to go for the perfect ten.

Never in his long and illustrious career had he even had three straight, let alone nine. But now the cards were falling his way, and with confidence raised to a fever pitch he decided to, as they say, go for it.

Mel was good, damn good. In fact, if records were ever kept, one might even say the best. He wasn't your usual run-of-the-mill obscene phone caller. No, siree. Mel had a style all his own. Most obscene phone callers try to shock young women with crass vulgar innuendos and explicit sexual language.

Mel used a different approach. He'd call up young women and ask them questions about math. He'd ask them how they'd like to solve a hundred logarithm problems or some horrible thing. Mel believed all women were mathophobes and had a thing about mathematics. More often than not after an Aaaaaggggggggg" came over the phone, he'd grin his reptilian grin and mark off another one for the gipper.

"Mason, Matson, Mattox, Maxwell, hmmm-mmm, Bugs Maxwell," he said to himself. "Just the untapped *mathaphobe* I'm looking for," he snickered to himself. He quickly dialed and waited for his prey to answer.

"Hello," said a soft voice at the other end.

"Hi, Bugs," he cunningly stalked his quarry.

"Who is this?"

"Say Bugs, how'd ya like me to tie you down and make you watch me solve the little `ol quadratic equation

$$x^2 + 2x + 1 = 0$$

"Who is this?"

"Just someone who's going to make you factor

$$mx - ny - nx + my$$

all night long." Mel's juices were starting to flow. He grinned to himself in his feral foxy way. He knew she was probably petrified and going into mathematical shock. Little did Mel know he was talking to a national finalist in the Putnam Exam.

"Well, I'd be glad to big boy, just after you multiply

$$(2x + y)(x + 2y)$$

came back through the phone.

KA-POW! Mel staggered back, taken completely off guard. He hadn't realized that women were getting into new disciplines. They were into professions which in the past were out of reach. Women were now entering the male domains of the physician, lawyer, politician, CEO, and even others not so well known, like the research mathematician.

"What's wrong tiger are you afraid of some little `ol algebra? How'd you like to integrate sin(x)," Bugs replied into the silent phone.

Mel was stunned, but he was a pro. He wasn't going to let any little cutie call his number. The bigger they are the harder they fall, he thought.

She probably learned a little math at girl scout camp.

"Bugs honey, I'm going to make you find common denominators all night long," he opened up a new area of mathematical exploration.

"Just as soon as you complete the square of

$$x^2 + 3x$$

Bugs shot back. *POW!* He staggered but still he didn't go down. He was a fighter and he cold roll with the punches the best of them.

"I'm going to differentiate exponentials all night long," Mel hit back with a sucker punch from calculus.

Be my guest," she said, "And I'll integrate the transcendental ones."

KER-BLAM! That sent Mel to the canvas. But still he got up and struck back, again she hit him, again and again. On and on they went, five minutes, ten minutes, twenty ...

By now the sparring was over, the gloves were taken off and the good stuff was coming out. These were not a couple of light-weights; they were two heavyweights in a knock-down-drag-out battle of the titans — a blow here, a blow there, harder, harder. *K-POW, GA-BONG, BA-VOOM, WOMPF, ZWOT, BOF, WAK, WOP, WOP, WOP.*

"I'm going to tie you down and make you solve logarithms," someone said from under the pile.

"And I'll tie you down and make you do exponentials," someone shot back.

WAK! BAM! P-THUNK! The fight to the death continued for thirty minutes, forty-five, one hour, two. *DOINK! BAM! BAP!* "I'm going to reduce all your fractions and simplify your binomials."

But anything repeated for the umpteenth time starts to lose its punch. Slowly, the punches began to lose their sting, way too weak for a knockout.

So what do you know about automorphic functions, dude?

"I'm going to solve $x - 1 = 0$," Mel swung for the fences. He was on the ropes fighting for his life.

————¤¤¤☒☒☒¤¤¤————

However, there was something that Mel was completely unaware. Bugs had been toying with him. She had suckered Mel with the old "rope-a-dope" strategy, leaning against the ropes, absorbing the best Mel could dish out, slowly, but surely, wearing him down. She was a cat toying with a mouse.

Mel was in the 15th round, punch drunk, hoping to land just one lucky punch.

"How'd ya like to add up ... Mel gasped in desperation.

Finally, it was time for Bugs to drop the hammer with a knockout punch she had been saving.

"Ok, pal, how'd ya like to solve

$$\frac{\partial^2 u}{\partial r^2} + \frac{2}{r}\frac{\partial u}{\partial r} + \frac{1}{r^2}\frac{\partial^2 u}{\partial \phi^2} + \frac{\cot\phi}{r^2}\frac{\partial u}{\partial \phi} + \frac{1}{r^2\sin^2\phi}\frac{\partial^2 u}{\partial \theta^2} = 0$$

Ka-Powie!! It hit Mel like a ton of bricks. He didn't even understand the equation, let alone solve it. The contest was over in a heartbeat. Mel just stood there, numb, ashen, glassy-eyed, unable to mutter a single filthy word. He was down for the count.

Bugs sat motionless, her ear pressed against a phone that just went dead. Finally, after a few minutes she slapped the phone back on the cradle.

"Well, what'd ya know," she thought, "another

math obscene phone caller TKO-ed by Laplace's equation."

$$\Pi\psi\Delta\Phi\theta\Lambda$$

6

HOW I (ALMOST) PROVED FERMAT'S LAST THEOREM

At the time I thought it rather strange. After all, any alien space jockey worth his plasma knows he's supposed to touch down in an alligator swamp in South Georgia and scare the holy bejesus out of some Georgia hillbillies. This time, however, they decided to land in the Maine woods and scare the holy be-jesus out of us Maine *woodsbillies*.

I was taking an evening stroll behind my house when suddenly I saw it, the unearthly green iridescent glow through the trees, the oval-shaped saucer hovering above the clearing. I hid behind some bushes and watched as a small opening materialized on the underbelly of the capsule and several otherworldly creatures, each no more than three-feet tall, emerged. Each had a human-like frame but with unmistakably alien features, a rubber-like skin that stretched over a bony scaffolding and two huge eyes that radiated an eerie green glow from a pasty-white face. I watched as no less than a dozen of these creatures made their way down a ladder. They all appeared similar except for one, which had a row of medals on its chest. I took it to be their leader.

After they reached the ground, the leader looked in my direction and shouted, "Farlow, we're here to make a deal." My body froze. "What's wrong, Farlow, is this the first time you've ever seen *Orkians* before?" the leader said as the creatures surrounded me. One of the creatures waved his bony fingers before my eyes and suddenly my fear was gone. "That's better," the leader said. "Now let's get down to business."

"Business?" I stammered.

"You're a mathematician, aren't you Farlow?" the creature said. "And by the way, we don't like to be thought of as creatures, you know we can read minds, we prefer aliens. My name's Harry."

"Yes, Harry," I said. "I'm a mathematician."

"And not a very good one, we understand. This makes you the perfect candidate," the little creature said. A sharp pain ran through my body.

"I warned you, Farlow. We have ways for dealing with thoughts like those."

"I won't let it happen again, Harry," I moaned, "but I'm an excellent mathematician."

"Don't give us that," Harry said. "We know you haven't made a single discovery in your entire mathematical career."

"What about my proof of the *Widdlestein Conjecture?*" I protested.

"You call that a discovery?" Harry laughed. "Our kids learn that in kindergarten. But don't worry Farlow, we're going to give you the answer to the most famous mathematical problem in the world. We're going to make you a star."

"Wow!" I started to like the little guys.

"Just a few gallons of plasma should do," Harry said.

"What?" I asked.

"One discovery for just a little plasma," Harry said, "You didn't think we'd give you the answer to the most famous mathematical problem in the world for nothing, did you? A couple of gallons should do it right guys?" The other aliens nodded in unison.

"Wait," I protested. But before I could say anything the alien with the bony fingers waved them before my eyes.

——¤¤¤ΞΞΞ¤¤¤——

The next thing I knew I was stretched out on a metallic table looking up at a small alien in a white lab coat. He was shining a beam of light in my face with a strange unearthly object which, for lack of a better description, looked like an ordinary flashlight. I panicked and screamed out.

"What's wrong, Farlow?" Harry asked. "Haven't you ever seen a flashlight before. Stop playing around with that damn thing, Raymond. And take that doctor's coat off. Damn kids," Harry said.

My body lie enmeshed in a tangle of wires and tubes. Ooze of several degrees of *yeeeeuuuuck* flowed like sea water through the tubes. I noticed that one of the tubes seemed to end somewhere in an opening in my abdomen. "What's all the gunk in that tube?" I asked.

"Oh don't mind that," Harry said. "It's just something we add."

"What about our deal?" I protested, trying to get up. "You said you were going to make me a famous mathematician."

"Yeah, yeah," Harry said. He then crossed the room to a filing cabinet and removed a large manila envelope from the top drawer. "Here it is," Harry said. "The answer to your most famous mathematical problem, a problem you earth people have been trying to solve for 358 years, a proof of Fermat's Last Theorem."

"What?!" I yelled, struggling to get up. "I'm giving you my plasma for Fermat's Last Theorem? That problem was solved 20 years ago!"

"What?" Harry seemed surprised.

"That's right," I said. "An Englishman proved it."

"Oh, sorry about that," Harry said. "We've been on the road so long you know. When we left they told us you didn't have a clue. What is that old theorem anyway?" Harry asked.

"I thought you were so smart," I said sarcastically.

"It's been a long time since kindergarten," Harry said.

"Well, let's start with something on that level then," I said. I thought it best if I gave Harry a beginner's lesson on the problem.

"You'll agree that

$$3^2 + 4^2 = 5^2$$

don't you?" I asked.

"Our kids ... ."

"Yeah, yeah," I interrupted. "They learn it in kindergarten"

I also told Harry there are other pairs of numbers whose sum of squares is the square of a third number, such as

$$5^2 + 12^2 = 13^2$$

"What does all this have to do with Fermat's Last Theorem?" Harry asked, starting to get bored. I told Harry that although there are other integers like the above that satisfy the equation, the French mathematician, Pierre de Fermat, claimed there weren't any non-zero ones that satisfied

$$a^n + b^n = c^n$$

when the exponent n is greater than 2, like 3, 4, ... Fermat scribbled this claim in the margin of a book but said his proof would not fit in the margin. And so for 358 years many of the world's greatest mathematicians tried without success to prove Fermat's claim, known as Fermat's Last Theorem.

Finally, in 1994 a 40-year old English mathematician from Princeton University, Andrew Wiles, solved the problem.

"If you would have come 20 years ago, I'd been famous," I yelled at Harry. "*Harry, where the devil are you?*"

I then realized the aliens weren't the slightest interested in Fermat's Last Theorem, and were all clustered around a huge apparatus. Suddenly, I heard a voice cry out, *"How do you like the new me, Farlow?"* I looked up and saw the most hideous looking alien I'd ever seen, it was horrible. *I then realized I was looking in a mirror!*

Harry and some aliens looking for plasma

"Aaaaaaaggggggghhhhhhhh," I screamed. WHAT HAVE YOU DONE TO ME?!" Looking around the room, I saw a dozen spitting images Brad Pitt.

"It's amazing what a little plasma will do for your complexion," one of the Brad Pitts said in a voice that sounded an awfully lot like Harry.

"Everyone wants to be a Brad Pitt, you'd think someone would pick a Clooney or a Redford."

"*Aaaaaagggggggggggghhhhhhhhh*," I screamed out again. "No, no, ... give me back my face, keep the damn theorem ... ."

——————¤¤¤ΞΞΞ¤¤¤——————

".... wake up dear, you're dreaming again," someone yelled at me. I found myself sitting in my own bed, drenched in sweat. My wife was shaking me.

I looked at her but questions remained. Had I been dreaming, or had I actually been visited by creatures, OUCH, aliens? If so, might they return and give me the answer to the Riemann Hypothesis or what about the Goldbach Conjecture? Only time would tell. I could still be a famous mathematician.

"Proving Fermat's Last Theorem again dear?" my wife asked wryly.

"Of course not," I said. "It's been proven, go back to sleep.

$$\Pi\psi\Delta\Phi\Theta\Lambda$$

7

FARMER BROWN'S REVENGE

If ever there was someone who had a well-founded bone to pick with the mathematics community, in particular all the mathematics textbook authors, it was a Robert M. Brown, a farmer from Plainville, Iowa.

"I couldn't turn around without some lamebrain textbook author noseying around the farm," he once told a reporter interviewing him over his life of harassment.

Mr. Brown's problems began back in the late 1800s when he wrote a letter to a local newspaper describing the best way to construct a fence around a hog lot which maximized the enclosed area at minimum cost. His solution was so ingenious he became well-known to area farmers as an important resource for solving mathematically-related problems which often arise on a farm. Over the years he was simply referred to as "Farmer Brown."

"Everything was going fine," Mr. Brown told the reporter, "until that damn textbook writer arrived and started asking stupid questions, like

"If a farmer had ducks and cows and if the farmer noticed the animals had a total of 12 heads and 32 feet, how many cows and ducks did the farmer have?"

"I told him I had two pitchforks, each with eight

prongs, and three scythes with long handles, and I asked him how many ways I could stick them up his ., well you get the picture," Farmer Brown told the reporter.

But unfortunately, Farmer Brown's warning went unheeded. When the textbook writer's new book, *Beginning Algebra with Farmer Brown,* came out, it contained dozens of word problems featuring Mr. Brown, always referring him as "Farmer Brown." A typical problem might read

"How should Farmer Brown construct a fence to contain his goats, how big should he should build his barn, or how many goats he should raise."

"I never raised a goat in my entire life," he told the reporter.

One of the popular Farmer Brown problems was the legendary "field-by-the-river" problem, which asks the following question.

Problem: Farmer Brown has 900 feet of fence and

wants to fence in as much area as possible for his wild turkeys. The fenced field is next to a river and should be rectangular in shape, but because of the river, there are only 3 sides that need to be fenced. What are the dimensions of the largest area Farmer Brown can construct to fence in his turkeys?

"Anyone building a fence for turkeys is as dumb as a corn cob," Mr. Brown told the reporter. They have wings, you know. And if they didn't, they can swim."

"My reputation was ruined," Mr. Brown told the reporter. "Those Farmer Brown problems made me look like a complete nitwit, wandering around the place with nothing better to do than solving simple word problems. I have a farm to run, you know."

Some books even included Mrs. Brown problems.

"If Mrs. Brown bakes two pies, and if ... ," a typical problem might begin.

"It's one thing to slander me, *but my wife?*" Mr. Brown told the reporter.

And if that wasn't enough of an indignity, Farmer Brown problems were generally written in a style that made them about as interesting as the history of, say, the plumbing-supply industry.

"I'll give ya a Farmer Brown problem, right up ... ."

"The name "Farmer Brown" gives kids more nightmares than Frankenstein," Mr. Brown told the reporter. "I can't leave my farm without little kids running behind, throwing sticks and singing Farmer Brown rhymes

Unfortunately, Mr. Brown died in the early 1900s with his reputation in shambles, always remembered as Farmer Brown from beginning algebra books, the obsessive farmer, forever building fences.

———¤¤¤ΞΞΞ¤¤¤———

One would think that would be the end of the Farmer Brown story, but not so. Recently, the heirs of Mr. Brown have filed a class action law suit against 35 publishers of mathematics textbooks, beginning algebra teachers, and everyone whose livelihood depends on the teaching of Farmer Brown problems. The lawsuit reads as follows:

———¤¤¤ΞΞΞ¤¤¤———

We, the descendents of Mr. Robert M. Brown, formerly of Plainsville, Iowa, demand a reparation in the sum of $150 million dollars from said publishing companies and mathematics teachers for the maligning of aforementioned Mr. Brown over the past 150 years by means of "Farmer Brown" problems. The portrayal of Mr. Brown as *a zealot farmer obsessed with solving meaningless problems.*

Upon Mr. Brown's heirs winning the $150 million dollar suit, one heir posed a final "Farmer Brown" problem for the publishing houses:

———¤¤¤⊟⊟⊟¤¤¤———

Problem: *Suppose the heirs of Farmer Brown sue publishing companies, Company A and Company B, for defaming the name of Mr. Brown (aka Farmer Brown). Suppose it costs $50K in lawyer fees to sue Company A and $75K to sue Company B. If settlement from Company A for each unauthorized book is $15 and the settlement for each unauthorized book from Company B is $15K, what should the heirs spend on each legal suit to maximize profits?*

———¤¤¤⊟⊟⊟¤¤¤———

"Now that's what I call the perfect Farmer Brown problem." remarked one of Farmer Brown's descendants.

Πψ∆ΦθΛ

8

3001: A CALCULATOR ODYSSEY

———

There was a knock on the door.

"Come in," Principal Whipple said looking up from his desk. The door opened and a youngish-looking man followed by a young girl entered the room.

"What can I do for you?" Principal Whipple said

"If you have a few minutes, we would like to talk to you about the mathematics curriculum here at Treadwheel Middle School. We would like to investigate the possibility of introducing some new ideas into Math 213," the older of the two said. He was Mr. Flanders, head of the Mathematics Department at the Treadwheel, who had brought with him his best mathematics student. The young girl had spent the summer working on a special project for Mr. Flanders and had discovered some exciting new ideas for doing mathematics.

"We're always ready for new ideas," Principal Whipple said. "Why don't you pull up some chairs," he continued. Whipple was skeptical about the whole affair, but didn't want to offend his young teacher.

Mr. Flanders began. "It's about pocket calculators," he said.

"What about `em?" Whipple asked.

"Well,... ," Mr. Flanders hesitated.

"Well?" the principal pressed.

"Well, my student, Bugs Maxwell, has found out a way to get around `em."

"What do you mean, 'around em'?" Whipple asked.

"Just that," Flanders said, "I assigned Miss. Maxwell a summer project to look into the history of the pocket calculator, and in the process, she discovered how to carry out all the arithmetic operations with only paper and pencil."

"That's a little hard to believe!" the principal replied. "Those calculations are quite difficult and have been wired into the calculator."

"Let me show you sir," the teacher responded, "what's six times eight?" Miss. Maxwell thought for a minute and then replied, "Forty-eight." The principal pulled his pocket calculator from his desk drawer, and after entering the numbers in his calculator, he pushed the multiplication key.

"By george, she's right," Whipple said.

"What's nine times four?" the principal then asked the girl. Miss Maxwell thought for a moment and then said, "Thirty-six." Again, Whipple checked the answer.

"Amazing," he said.

"She can even multiply large numbers," the teacher said.

"No!?"

"Multiply 112 times 63," Mr. Flanders instructed the girl. This time the girl took out a sheet of paper and wrote down the numbers. A few minutes later the girl was finished.

"7056," she announced.

Whipple quickly checked the answer on his calculator. "I don't believe it," he said. "How in the world does she do it?"

"Like I said," the teacher spoke, "I assigned her a project to look into the history of the pocket calculator, and in the process she found some old manuals that explained how calculators actually did their calculations. She has been practicing those old rules all summer. Not only can she multiply, but she can also add, subtract, and even divide."

How the devil did she do that?

"*Divide!*" Whipple gasped.

"The rules are actually fairly simple. I have been practicing them myself for the past month and already I can add and subtract," the teacher said.

"Amazing," Whipple said.

"This brings us to the reason why we wanted to talk with you," the teacher continued. "We wanted to introduce hand computation into Math 213. We could teach the students how to add, subtract, multiply, and divide by the time they graduate from high school."

"It would be a fantastic accomplishment if it could be done!" Whipple said. "But what's the point?"

"But don't you see," Flanders said, "by learning these basic operations, the student would learn to think and be freed from the routine tasks of operating and maintaining calculators, computers, and an assortment of electronic devices. The student's mind could be used for more creative tasks."

"The students could even learn how to compute areas of squares, rectangles, triangles, and other geometrical shapes without resorting to their calculators," the teacher went on. "It is possible that in the future, the calculator might not be needed at all."

"Incredible!" Whipple thought, absolutely

incredible. "*There is no limit to the human mind,*" *he was thinking.*

$$\Pi\psi\Delta\Phi\Theta\Lambda$$

9

THE YIN-YANG OF MATHEMATICS AND POETRY

———

One little bee flew and flew.
He met a friend, and that made two.
Two little bees, busy as could be,
Along came another and that made three.
(and so on)

You see what's going on, don't you? By the time

———

we reach the advanced age of three, we are already learning to count by rhyming words, in this case about five little *Apis mellifera* in the nursery rhyme, *Five Little Bees*. For a child the hypnotic rhythm of rhyming words in story form delights the child, resulting in learning and retention of the (arguably, less than delightful) abstract subject of numbers and counting.

Grade school students in past times were aware of poems such as

> *Six times eight fell off the plate, that's what makes it forty-eight.*

The use of poetry as an aid in learning mathematics is not limited to preschoolers. In some cultures, notably India in the Middle Ages, mathematical knowledge is often imparted by rhymed verses.

But the connection between poetry and mathematics is not a one-way street. Who among us cannot see the poetic beauty in the symmetry and precision of the algebraic expressions:

$$a$$

$$a + b$$

$$a^2 + 2ab + b^2$$

$$a^3 + 3a^2b + 3ab^2 + b^3$$

or are humbled by the equation

$$e^{i\pi} + 1 = 0$$

first expressed by the mathematical laureate, Leonhard Euler (oi'-ler). The equation links the five most famous numbers of mathematics in a single equation? Maybe it's a proof of the existence of God. Maybe we've always known the *answer*, we just didn't know the *question*.

———¤¤¤⌕⌕⌕¤¤¤———

It has often been said that mathematics and poetry are our yin and yang, apparent opposites on the outside, but complementary on the inside. This philosophical observation is supported by the philosopher, Thomas Hill, who said, although mathematics and poetry both appeal to the imagination, mathematics appeals to the intellect while poetry is addressed to the heart. This philosophy is further supported by the writer William James, who wrote

"The union of the mathematician and the poet, passion with correctness, is surely the ideal."

Further, there are mathematicians who believe poetry is crucial for the development of the mathematical mind. No less than the "father of modern mathematical rigor," the 19th-century German mathematician, Karl Weierstrass, once said

"A mathematician who is not somewhat of a poet will never be a perfect mathematician."

Although you may be rusty on the Weierstrass definition of continuity, are you not able to see the poetic flair in his mathematical definition

$$(\forall \varepsilon > 0)(\exists \delta > 0)(\,|x - a| < \delta \Rightarrow |f(x) - f(a)| < \varepsilon\,)$$

or the Euler-Lagrange equation

$$\frac{\partial L}{\partial q} = \frac{d}{dt}\left(\frac{\partial L}{\partial p}\right)$$

which shows how physical systems evolve over time. The English writer G. J. Chesterton once observed

> "The difference between a poet and a mathematician is that the poet tries to get his head into the heavens while the mathematician tries to get heaven into his head."

The poet Sir Waldo Emerson went even further when he expressed relations between poetry and mathematics when he once said

> We do not listen with the best regard to the verses of a person who is only a poet, nor to his problems if he is only an algebraist; but if a person is acquainted with the geometric foundations of things and with their festal splendor, his poetry is precise and his arithmetic musical."

Although there are some who believe the poetic mind plays a role in the formulation of mathematical thoughts, others claim the converse is true, that mathematical thought plays an important role in the development of a poet. After all, does the poet not apply mathematics in writing a precise cadence of *iambic meter?* And does not the writer of the sonnet use mathematical precision to embody age-old emotions with a minimal number of lines? And is it not true that a poet's verses display mathematical symmetries? Longfellow was aware of the mathematical principles of symmetry and precision, when he wrote the last stanza of *A Day is Done:*

> *And the night shall be filled with music,*
> *And the cares that infest the day,*
> *Shall fold their tents like the Arabs,*
> *And silently steal away.*

The succinctness of both mathematics and poetry, observed by the English philosopher Francis Bacon, lies in the fact that both disciplines seek only the truth, as Bacon states

> *"No pleasure is comparable to the standing upon the vantage ground of truth."*

On one hand, poetry seeks universal truths, not directly from words, but by circuitously using images formed from words. On the other hand, mathematics seeks the truth by means of symbols. The union of mathematics and poetry, passion with correctness, is an ideal we cannot dismiss. But of all the bonds that link mathematics and poetry, possibly the strongest is the endless invention of each discipline. The English poet Samuel Johnson once remarked

> *"The essence of poetry is invention, of introducing something unexpected, which surprises and delights."*

The same could be said of mathematics. Neither mathematics nor poetry has a boundary that marks an end of its domain. The only boundary lies in the imaginations of its practitioners.

$$\Pi\psi\Delta\Phi\theta\Lambda$$

10

GOD EQUATION: SENTRY AT THE PEARLY GATES

Wrong!" the old woman at the gate cackled fiendishly in a voice that could raise the dead, which considering the fact that everyone within earshot *was* dead, made the entire line jump to attention. The meek-looking man who was the object of her ridicule lowered his head and slunk off.

"Next," the old bag hissed. I had finally reached the front of the line, whereupon I approached the gatekeeper and gave her my manila envelope. She tore it open and began rifling through the papers. "Farlow?" she said looking over a pair of antiquated spectacles. "Is that *you*, little Jerry Farlow?

I looked into the beady eyes that peered from the face of this sorceress. Suddenly something inside clicked.

"Ms Hammerschnozle?" I asked. "Is that you?"

"Of course it's me. *But what are you doing here?*"

"We all die sometime, Ms. Hammerschnozle," I said. "I didn't want to go to the other place. But what are you doing here?"

"*Where did you think I'd be?*" she said snidely.

"Oh, I knew you'd be here," I lied through my teeth.

"But isn't Saint Peter supposed to be a *male* angel."

"Well, as usual Farlow, you thought wrong," The old bag said. "Times have changed. They finally got some religion up here and brought in someone that knew a thing or two about math."

God! My ultimate worst nightmare! My infamous tenth-grade math teacher, Ms. Hammerschnozle, whose endless aspersions on my academic reputation almost a century before, was now a stand-in at the Pearly Gates for no other than, alas, Saint Peter. And she would determine whether my final destination would be the `ol harp farm in the sky or a permanent reservation along the River Styx.

"You know," I smiled at Ms. Hammerschnozle. "I've learned just a little mathematics myself since my school days. Why don't you just give me the test." I smiled smugly, knowing there was no way she could ask me something about mathematics I didn't know.

"Ok, Farlow," she said. "What do you know about the *God Equation*."

"*The God Equation?*" I croaked. A river of sweat

the size of the Ganges ran down my face. "Uh, I don't think I'm familiar with that particular equation."

"It's God's own equation," she said. "This single equation contains the most famous numbers in all mathematics. If you can answer my questions about these numbers, you will be allowed to enter Heaven. If not you will ... ."

"Yes, yes, I understand," I shuddered to think about the alternative.

"Are you ready, Farlow?" she asked. "The first number is about the number 1. What do you know about 1?"

The Number 1

"Was she kidding?" I chuckled to myself. *Eins, Zwei* good `ol Numero Uno, the first of what we call the *natural numbers*

$$1, \ 2, \ 3, \ 4, \ ...$$

In addition to being the first natural number, it is the only number that has the property that the product is unchanged when multiplied by other numbers. For example

$$1 \times 6 = 6 \quad 1 \times 34 = 34 \quad 1 \times 68 = 68$$

"Ok," Hammerschnozle interrupted, "Enough of the kid stuff. I think we can get on to something more challenging." I waited for the guillotine to drop. "What can you tell me about π, the most famous number in geometry?"

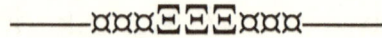

The Number π

I chuckled to myself. Although π isn't an integer like 1, it is one of the most famous numbers in mathematics and was well known by Greek mathematicians over two thousand years ago. Quite simply, it is the ratio of the circumference of a circle to its diameter. The first eight digits of π are 3.1415926, which means that the circumference of a circle is slightly more than three times its diameter.

"Ok Farlow, you can stop with all that," Hammerschnozle interrupted. "You're halfway home. Are you ready for the third great number of mathematics?" she asked. "Each number gets just a bit harder."

I started to squirm a little. "Yes," I finally said.

"Ok," she said. "What can you tell me about the imaginary number *i* ?"

The Imaginary Number *i*

I was now on a roll. The imaginary number *i* is one of the most fascinating numbers in all mathematics. You might say the origin of the number *i* began in the 16th century when Italian mathematicians Gerolamo Cardano and Niccolo Tartaglia tried to solve 3rd and 4th order polynomial equations. In the process they ran into the square root of -1, or $\sqrt{-1}$. They considered the square root of a negative number an impossibility and so it was taken as a 'non number,' but after some time mathematicians found it useful and so it was accepted as a number, although sometimes called an imaginary number, which we denote by *i*. The entire subject of complex numbers has applications in ...

"Ok, ok, you can stop now," Ms. Hammer-schnozle interrupted. "It's clear you know a little about mathematics. You told me about the three famous numbers 1, π. and *i*. We now have only one more number to go but it is the most difficult. What can you tell me about the number *e* ?"

———¤¤¤ΞΞ¤¤¤———

The Elusive Number *e*

"What?" I thought, no trick question? Every mathematician worth his white board knows about *e*. In some areas of mathematics, like differential equations, it is arguably the most famous number of all. The number *e* is a real number like the numbers 1 and π. However, it is more difficult to describe since it is generally defined as the limiting value of some other numbers. The Swiss mathematician, Leonhard Euler, defined *e* as the limiting value of the expression

$$(1 + 1/n)^n$$

as *n* gets larger and larger. It is used by engineers to describe growth and decay of ...

"That's enough," Hammerschnozle said at last, "No more sugar coating the questions."

Math exam at the Pearly Gates

I looked over her shoulder to see if the Pearly Gates had started to open. "Now just tell me about the God equation and we're finished," she said.

"*What?*" I said.

"Suppose we combine 1, π, *i and* e into the single quantity

$$e^{i\pi} + 1$$

she said. "What new number do we get?"

"*It's impossible,*" I yelled. Just the thought of combining these four numbers into the single number gave me chills.

"Wrong!" she barked. The value of $e^{i\pi} + 1$ can easily be found. *In fact it's zero!* The great Swiss mathematician Leonhard Euler used his famous equation

$$e^{ix} = \cos x + i \sin x$$

and simply plugging in *x* = π, he got

$$e^{i\pi} = \cos \pi + i \sin \pi = -1 + i\,(0) = -1$$

thus

$$e^{i\pi} + 1 = -1 + 1 = 0$$

"Up here we call it the *God Equation*," Hammer-schnozle said.

"*What?*" I gasped in disbelief.

She continued, "This single equation brings together the five most famous numbers in all mathematics; the basic whole numbers 0 and 1 from arithmetic, the fundamental constant π of geometry, the complex number i, and the core constant e of calculus."

Next!" the old bag cackled heinously.

"*Aaaaaaaaaaaahhhhhhhhhhhh,*" I screamed as two burly angels grabbed me by my arms. "No, no," I protested. "I belong here. Give me another chance."

"*Wake up, wake up,*" I felt someone tugging on my arm. I was sitting up in bed, drenched in sweat, looking at my wife. "Those nasty little monkeys carrying you off again, Toto?" she said.

"Would you believe it? Did you know

$$e^{i\pi} + 1 = 0 \text{ ?}''$$

"Yeah, yeah" she said rolling over, "The God Equation. Go back to sleep."

$$\Pi\psi\Delta\Phi\Theta\Lambda$$

II

IN VIRTUAL REALITY, #NERDSRULE !

———

The following story contains mature and graphic subject matter: reader discretion is advised.

———¤¤¤ΞΞΞ¤¤¤———

I pressed her creamy bare softness against my chiseled torso and while she whimpered like a puppy, I kissed her ripened lips. "Take me, take me," she cried out in helpless surrender, unable to contain herself. Just another math groupie, I was thinking.

———¤¤¤ΞΞΞ¤¤¤———

Oops, excuse me, I was just channeling my inner nerdy-ness and got a little carried away. Just go back to what you were doing … .

But things are changing for us mathy types. The hens have come home to roost, it's revenge of the geeks. What goes around comes around. He who laughs last, laughs best. *Heh, heh, heh.*

Talk about your geek revenge. I couldn't help stop laughing when I read in *Modern Computer News* that computer software companies are sucking up to computer graphics programmers, begging them to write computer programs for virtual reality systems. With the dawning of the digital age, everyone from gaming companies, the military, industry, and even Hollywood producers are rummaging through computer science departments, seeking out the latest crop of computer graphics graduates.

"It's a digital gold rush out there," one airline executive said.

As for myself, I've waited a lifetime for the arrival of virtual reality systems. You see, when you talk about virtual reality, you're talking my language. Mathematics!

That's right pal, once the world has passed through the looking glass from the real world into virtual reality, you've arrived at my hashtag microcosm of #nerdsrule! It's just a bunch of o's and 1's. We're talking my kind of language; Boolean algebra, homogeneous coordinates, Bezier splines,.fracttal geometry!

Heh, heh, heh. I've got it now, baby. I'll control the world. Heh, heh, heh. Come to papa. Who's your daddy? Heh, heh, heh.

But there are more uses of virtual reality than modeling architectural designs, medical training and simulation, and virtual business presentations. It's cybersex! Cyberbabes! Cyberhunks! That's right, friend, you strap on a pair of virtual reality goggles and, voila, you're in virtual heaven, entertained by realistic, computer-drawn hunks and babes. And you won't be watching Snow White and the Seven Dwarfs, brother (well, actually you might now that I think about it). Why, as one adult software company executive said in a magazine article recently, "It is better than the real thing." (Hey, don't look at me, I'm just quoting the magazine.) But, whatever you're watching, everything is converted to 0's and 1's before transmitted.

In other words, don't feel guilty, *you're doing mathematics!*

$$\Pi\psi\Delta\Phi\Theta\Lambda$$

12

LISTEN UP, I'M NOT GOING TO REPEAT THIS

I'm starting to get annoyed with some of you. One would think that after all those endless hours in beginning algebra, you would have the technique of "simplifying an algebraic equation" mastered by now. But no doubt there are some among you who just don't get it. So pay attention, For that reason, I would like to give a short tutorial on how to simplify a mathematical equation. I'm not going to repeat this so stay alert.

Suppose you want to simplify the basic algebraic expression

$$y = x - x^2$$

As some of you probably learned in your algebra class. it is already in simplest form, but you still have a lot to learn. A few of you smarter ones will no doubt factor the right-hand side as

$$y = x\left(1 - x\right)$$

and choose this for your answer. Well, *au contraire*. We all agree the two above equations are equivalent and that the second one might seem simpler to some, but it is *still* possible to go a little further if you use a little imagination.

We begin by taking the reciprocal of each side of the last equation, getting

$$\frac{1}{y} = \frac{1}{x(1-x)}$$

and if we carry out the division of the ratio 1/(1-x) which appears on the right-hand side, we get

$$\frac{1}{y} = \frac{1}{x(1-x)} = \frac{1}{x}\left(1 + x + x^2 + \cdots\right)$$

We can now multiply each side of this equation by the expression

$$\frac{e^{-x^2}\cosh(\pi x)}{\sqrt{1+x^2}}$$

and integrating from 0 to z, yields the resulting equation

$$\int_0^z k(x)\left(\frac{e^{-x^2}\cosh(\pi x)}{\sqrt{1+x^2}}\right)\,dx$$

where of course

$$k(x) = \frac{1}{x} + 1 + x + x^2 + \cdots$$

The above integral on the right is now easily recognizable as the *Kleinhopper Integral* and so by integrating again and solving for y, we get the final form

$$y = \iiint\limits_{\Omega} \sum_{k=1}^{\infty} \sum_{j=k}^{\infty} \left(\frac{\partial^2}{\partial x \partial y} \left(A_{jk}^{\alpha\beta} \frac{\partial u}{\partial x} \right) \right) + \int\limits_{\Delta} \left(\sum_{k=1}^{\infty} a_k e^{-(k\lambda x)^2 x} dx \right)$$

Of course, this equation could be written in oblate spherical coordinates, *but what would be the point?*

$$\Pi\psi\Delta\Phi\theta\Lambda$$

13

AN OLD COLLEGE TEXTBOOK YOU MIGHT RECALL

———

Sometimes I wake up at night drenched in sweat. Far off, I hear faint cries of tormented students. I am well aware of anguish endured by hundreds upon hundreds of these tortured souls. Then, it strikes me, maybe they know the whereabouts of where I lie, and now their cries grow louder and louder. They are coming for me by the hundreds, thousands, ... My heart is pounding like a drum. They're getting CLOSER, CLOSER! Then I think of the massive royalties that keep rolling in, and I turn over and sleep like a baby.

One thing I've always eschewed, well, maybe not eschewed, but avoided like the plague, is the fact that I might be the person that wrote those overpriced textbooks that sucked your bank account dry back in your college days. You know, all that hard-earned cash you expected to go for booze and wild parties, went instead to buy my seminal text, Adventures with the Quadratic or my all-time favorite, A Passion for the Polynomial..

It gives me no pleasure knowing that at any hour of the day there are thousands of poor college freshmen using my name in vain.

"What in the hell is this guy talking about?" is probably running through the minds of countless numbers at this very moment. Or no doubt "BORING," is also a common theme. And, of course, the always popular, "Where the hell are the answers to the problems?" That, of course, is a dumb question. They should know by now if they want the answers to the problems, it'll cost them an extra $59.95. Did they actually think the answers came with the textbook? *Ha, ha, ha ,... .*

Why did I put my address in that book?

I have been called the Stephen King of Algebra. Students experience more terror on the first page of my *Adventures with the Quadratic* than in a Stephen King novel. For that reason, I would like to share with you a few secrets I've developed over the years.

Rule 1: The Used Book Paradox

If there's one thing of which I take umbrage, it's the humongous line of students that form at the college bookstore at the end of the semester. They, of course, are carrying out the unspeakable act of *reselling* my books. Although the extra money may come in handy, I contend that keeping the textbook has several benefits. First, the student is able to reread the book at a later date, developing new insights. Secondly, the student adds another book to his mathematical library. And third, although not important but I mention it anyway, it knocks a hole in my royalty check big enough to drive a Mack truck.

To help the beginning student avoid the cheap lure of the resale counter, I have devised several effective strategies. One particularly effective method is my design of an ornate book plate, embellished on the cover of the book where the students can sign their name, campus address, and other relevant information. This information will be noticed by anyone finding the book if accidentally misplaced. It will also be noticed by the bookstore manager, classifying it as worthless junk.

—————¤¤¤ΞΞΞ¤¤¤—————

Rule 2: Textbook Supplements

To assist schools who adopt my books, I provide instructors with a long line of pedagogical "extras," which includes *sample tests, computer software, instructor manuals, solution manuals, teaching guidelines, sample syllabi,* ... after which the school will have become so enmeshed in a maze of supplements that to change the book would require the skills of a licensed accountant.

—————¤¤¤ΞΞΞ¤¤¤—————

Rule 3: Proper Font Size

A major expense of producing a new book is the cost of paper. For that reason, I avoid printing equations, symbols and tables in excessively large font. There is little reason to use the blatantly large type for the equation

$$y = \frac{3x^4}{z} + \tan^{-1}x + \frac{1}{(e^z + e^{-z})} + \sin y + 2$$

when the perfectly smaller size

$$y = \frac{3x^4}{2} + \tan^{-1} x + \frac{1}{\left(e^x + e^{-x}\right)} + \sin y + 2$$

will suffice.

$$\Pi\psi\Delta\Phi\Theta\Lambda$$

14

YOU DO THE CALCULUS: WE'LL DO THE CAT FOOD

Well, that was awkward!

Back when the earth was in its early cooling stage and I was in the early professor stage, I was assigned to teach a beginning course in calculus. Excited about my new charge as mentor of a new crop of eager learners, I resolved to impress upon the class the myriad of ways calculus is used in their daily lives.

One of the important topics in a first-semester course in calculus is an area called "optimization," where one uses the concept of the derivative to, well, find the optimal way of doing things. For example, suppose you find yourself on one side of a river and you want to swim to the point on the other side directly opposite you. Now, suppose the river is a mile across and flows 2 miles per hour and you can walk 3 miles per hour. The problem is how can you reach your destination in minimum time by a combination of walking and swimming?

Hard pressed to find a practical application to the "river problem," I decided to look for another.

The problem that caught my fancy was the "can problem," which seeks the dimensions of a cylinder with a given volume that has the smallest surface area (top, bottom, sides). If the can is too tall or too short relative to its width, the surface area is large with a given volume. It turns out the "minimum-surface area can" is the one with its diameter on the top or bottom equal to its height. The following diagram illustrates three different sized cans with the same volume V, but different surface areas.

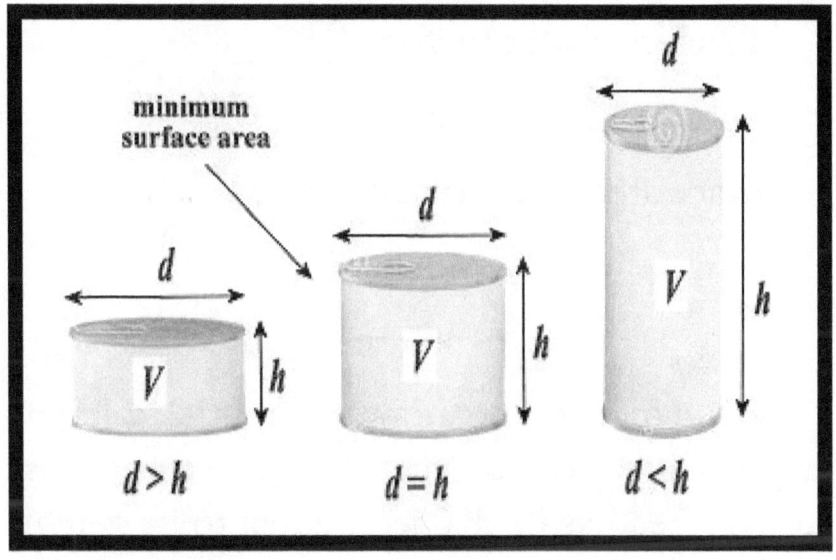

Cans with equal volume but with different surface areas

Upon observing that cat food is packaged in short, squatty cans with larger diameter than height, hence suboptimal insofar as minimizing the amount of tin required in making the cans, I rushed off a letter to the Carnation Company of Los Angeles, the company that makes Friskies

Buffet Poultry Platter, the favorite of my cat Dandy.

In my letter, I pointed out the relevant dimensions of a can of Friskies Buffet.

- diameter: $d = 3.2$ inches

- height: $h = 1.75$ inches

- surface area = 34.5 square inches

- volume = 14.5 cubic inches

I then demonstrated how differential calculus could be used to show that the can with smallest surface area has its diameter equal to its height, and for a can whose volume was 14.5 cubic inches, similar to a can of Friskies Buffet, the dimensions of the can would be as follows:

- diameter: $d = 2.65$ inches

- height: $h = 2.65$ inches

- surface area: $A = 2\,(1/2)\pi d^2 + \pi dh = 33$ sq inches

- volume: $V = (1/2)\,\pi d^2 h = 14.5$ cubic inches

In other words, the "minimum surface-area can" has a surface area of (approximately) 33 square inches, a decrease of 1.5 square inches from the 34.5 square inches of the can used by the Carnation Company to produce a can of Friskies Buffet Poultry Platter.

Author and Dandy watching Baywatch

I suggested to the Carnation Company that if they made cans with diameter equal to their height, the savings might be substantial. I suggested they might consider the services of a professional mathematician on the production floor.

A few months later, I received the following letter in the mail.

S.J. Farlow
Professor of Mathematics

Dear Professor Farlow,

We appreciate the interest you expressed in examining the height-to-diameter ratio of our food products. A 1×1 ratio of height versus diameter is the most efficient use of material; however, there are other factors which must be considered in designing a can for a particular product.

1. **Thermal Processing:** There is an inverse relationship between the most efficient design for cans relative surface area and the amount of processing time to sterilize the product contained within. In other words, a tall can or a short wide can will require considerably less processing time and energy to achieve commercial sterility than a can which is nearly equal in height and diameter.

2. **Warehouse and Shipping Efficiency:** Smaller diameter cans make more efficient use of packaging and shipping space.

As you can see, cost and efficiency of a container are related to factors other than just the amount of material used. These are only a few of the factors which must be taken into consideration when designing a can.

Sincerely,

Assistant Product Manager
Friskies Buffet

In other words, from the point of view of minimizing the time it takes to *sterilize* the contents of a can, the "diameter equal height can" is the absolute *worst* possible can!

The letter from the Carnation Company was their way of saying, thank you professor, but you can now return to your calculus class and we can return to making cat food.

$$\Pi\psi\Delta\Phi\theta\Lambda$$

15

PARDON ME MADAM BUT YOU JUST DRANK THE ORBIT OF JUPITER

———

To repay you for blowing all your hard-earned cash on this overpriced book, the author would like to disclose a tip on how to be the life of a party. After the party gets rolling and just when you are ready to savor a sip of your favorite *Sparkling Rose'*, look thoughtfully into the glass and announce to everyone within earshot that you are having an out-of-body experience, looking directly at the orbit of Mars.

If you are drinking wine from a glass like the (nearly cylindrical) one shown below, the surface of wine forms a circle when the glass is held upright with no tilt, but when the glass is tilted, the surface has the shape of an ellipse.

Since the orbits of all the planets are ellipses with various eccentricities (i.e. degrees of oblong-ness), the obvious question is how many degrees must someone tilt the glass to see the orbits of all the planets, or at least miniature versions.

Before continuing with our party activities, we digress for a brief review of ellipses. Roughly, an ellipse is an oblong circle, where the amount of oblong-ness, or how much it deviates from a circle, is called the *eccentricity*. There are formulae for the eccentricity of an ellipse, but we need not worry about those here.

untilted   tilted

Untilted and tilted wine glass

The circle in a special ellipse with eccentricity 0, and the more oblong the ellipse, the closer the eccentricity gets to 1. The eccentricity of an ellipse never actually reaches 1 since when it does, it's no longer an ellipse, but a parabola, and when the eccentricity is greater than 1, the figure is a hyperbola. The following figure shows ellipses with different eccentricities, which we denote by the letter e.

The orbit of the earth, although an ellipse, appears as a circle to the naked eye, having an ellipse of only $e = 0.0167$. On the other hand,

Halley's comet has an eccentricity $e = 0.97$, meaning that it is very elongated.

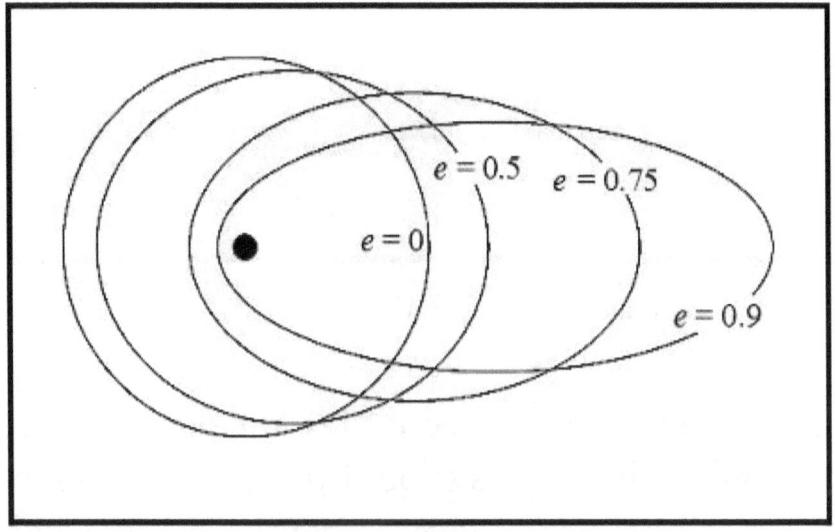

Figure 1: Ellipses with different eccentricities

Halley's Comet orbits the sun every 76 years and last passed by the sun in 1986, meaning hang around and you can see it in 2061. The drawing in Figure 2 illustrating the location of Halley's Comet as a function of time was drawn by Steven Dutch, Natural and Applied Sciences, the University of Wisconsin at Green Bay.

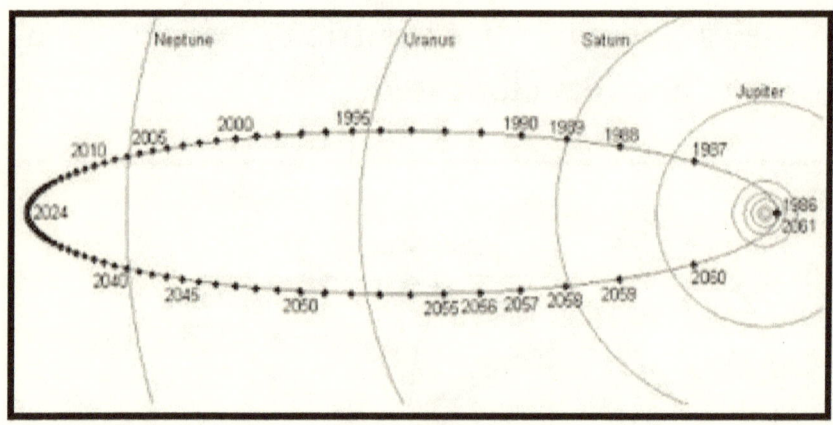

Figure 2: 76-Year orbit of Halley's Comet

Getting back to party activities, the big question is, how many degrees should you tilt the glass to see an ellipse with a given eccentricity? We will not carry out the freshman-calculus math here but if one tilts the glass θ degrees then the surface of the wine inside the glass forms an ellipse with eccentricity *e* given by the trigonometric formula

$$e = \sin \theta$$

where sin θ, as students of trigonometric know, is the sine function of the angle θ. This equation can then be solved for θ, getting

$$\theta = \arcsin (e)$$

where arcsin (e) is the inverse sine function of e.

If the reader would like to evaluate the inverse sine function of $e = 0.25$, which is the eccentricity of Pluto, just mosey on over to the website www.wolframalpha.com and key in arcsin(0.25) in the little box and you will get $\theta = 14.18$ degrees, the tilt angle required to see the orbit of Pluto.

For your viewing pleasure, the following table shows the required angle one should tilt a glass to see the orbits of planets. Yes, we know 'ol Pluto has been sent to the minor leagues and is no longer considered a bona fide planet, but we felt sorry for it so we included it here. Plus it has the largest angle of tilt of the other eight legitimate planets, comets and asteroids in our solar system.

If you tilt the glass one degree (0.96 degrees to be exact), you will see a miniature version of the orbit of the earth, which of course, unless you have eyes of a hawk. looks like a circle. You can then continue your expedition of the solar system by tilting the glass 5.36 degrees and view a miniature orbit of Mars (which still looks like a circle), but when you tilt the glass 14 degrees, you should be able to distinguish the orbit from a circle and see a scaled-down version of the orbit of (the ex-planet) Pluto.

Planet	Eccentricity	Angle of Tilt
Mercury	0.206	11.54 degrees
Venus	0.0068	0.39 degrees
Earth	0.0167	0.96 degrees
Mars	0.0934	5.36 degrees
Jupiter	0.0485	2.78 degrees
Saturn	0.0556	3.19 degrees
Uranus	0.0472	2.71 degrees
Neptune	0.0086	0.49 degrees

Asteroid/Comets	Eccentricity	Angle of Tilt
Ceres	0.078	4 degrees
Hidalgo	0.66	41 degrees
Halley	0.97	76 degrees
Kohoutek	0.999	89 degrees

The granddaddy of elongated-ness is the comet Kohoutek, which passed by the sun in 1973 and will return in 75,000 years. Its orbit has an eccentricity of $e = 0.999$, meaning to see it you must tilt the glass

$$\theta = \arcsin(0.999) = 89^0$$

degrees, which gives you an excuse to give your host a toast and partake the remainder of the wine.

$$\Pi\psi\Delta\Phi\theta\Lambda$$

16

STUDENT EXCUSES AND YADDA, YADDA, YADDA FROM THE CRYPT

————

The phrase "Tiger Mother" as introduced in Amy Chua's child-rearing story *Battle Hymn of the Tiger Mother* describes a mother's exercise in severe parenting. As for myself, while I was never an over-bearing mother, it was a campus cliché that I was the softest touch in academia.

So it came as a bolt out of the blue that when one of my calculus students came rushing hell-bent into my office, complaining about not understanding my lecture which I so clearly presented, and had the gall to call me a *Tiger Professor*.

Her painful charge cut me to the bone, but the wound was temporary and the moment it subsided I simply regained my composure and politely told her to *shut up and work on her derivatives till she got them all right*.

I only relate the incident since if I actually *had* given her such an instruction, she'd probably interpret it as an outrageous joke and would keel over and roll on the floor in laughter. You see, nowadays, the ancient master/slave model for the teacher/student relation is an artifact of ancient history. Nowadays, it's a more level-playing field when it comes to student "game-playing" with the professor.

:Anyone who has taught college students for more than ten minutes has been besieged by a myriad of excuses for missing exams, not handing in homework, or dozens of other undertakings the professor was crazy enough to assign. Those old canards of "The dog ate my homework" or "To tell you the honest truth .... ," are passe' in today's world of student smokescreens, where the unimaginative falsehoods of the past have been superseded by masterpieces of doublespeak.

I'll admit I've been guilty myself of inventing excuses, alibis, and overall fish stories for avoiding social functions. But my archive of evasions for cutting out of undesirable functions is child's play compared to those foisted upon me by students over the years.

My all-time favorite student hokum came from an all-conference linebacker on the university football team who seemed to be present on every tackle of an opponent's fullback, but not so much in my math class. I was under the impression he had dropped the class since he was invisible after the first day of class, both for lectures and exams, but on the last day of class he appears in my office and asks, "So, *where we go from here?*"

One thing is axiomatic. A student will never say, "I missed the exam because I was at the Tipsy Cow last night with the guys and you know how that goes." Although carousing at local student hangouts is not an excuse that will dupe your average professor, some students bring out the big guns and resort to family illnesses, even death. Most medical emergency excuses are genuine, but some, well, ...

One suspected medical con was fabricated by a student who asked if she could skip an exam to attend a grandfather's funeral. I naturally agreed to her request. A couple years later in another class she made the same appeal. I agreed again but it did cross my mind how I'd respond if she made the same request again.

Then there are students who are chronic favor-askers. I once had a student who always sat attentively during the entirety of my mesmerizing lectures, but often asked for bathroom breaks during exams. I could hardly deny his request and I sure as hell wasn't going to follow him into the men's room. I always boasted that my exams were taxing, but never thought they rose to the level of disrupting one's gastroin-testinal track. I will confess, however, I thought about checking out the men's room to see if any math equations were scrawled across toilet-stall walls.

———¤¤¤☒☒☒¤¤¤———

Another facet of the teacher-student dynamic is the manner in which a student addresses the professor. When I started my professorial duties fifty years ago I was generally addressed as "pro-fessor," but gradually that opening changed to "Hey," or simply, "Is this stuff going to be on the exam?" I never felt put-down by the less-than traditional academic address, but I did have a student who always alluded to me as "Sport." She was a member of the school field-hockey team, so I am assumed it was a term of endear-ment.

———¤¤¤☒☒☒¤¤¤———

If a professor is male and young, there is always the underlying current of non-professional rela-tionships with female students. When I was a first-year professor and recently married, my new wife, who at the time was an undergraduate, suggested that she attend my very first university lecture. Just to get an idea of the kind of mathematician she married, she joked. After class she made the wry observation that she might have to continue with the class since two female students sitting next to her commented on the "cute professor," so she said.

Then, when Christmas rolled around that first year one of the floors of a women's dorm asked if I would judge their Christmas decorations. My wife suggested such a task might require a "woman's touch" and offered to help me with the chore.

As the years passed, I noticed an absence of the occasional smile from a passing female student or a faint raise of a brow, and went about campus invisible to the female student body.

Well, at least until recently. Nowadays, the attention is back. A student, male or female, will sometimes offer to carry my books and academic gear to class. Recently, on my way to class and crossing an icy street, two of my female students took me by the arm and asked, "Professor, do you need any help crossing the street?"

At least they didn't call me Sport.

However, sometimes in the teacher/student relationship, it is the teacher who is delinquent. I once knew a teacher who constantly berated his students for any failing, however minor. He would stand before the class and give them the proverbial teacher warning that if the boys didn't study harder, they'd end up digging ditches the rest of their lives.

Many years later the teacher happened upon one of his former students at a school reunion. The student greeted his old mentor warmly and told the teacher's that the teacher's prophecy of not studying and spending his life digging ditches came true. The student went on to say *didn't* study and is now spending his life digging ditches — as the owner and CEO of the largest dredging company east of the Mississippi.

ΠψΔΦθΛ

17

EUCLID, GAUSS, AND FERMAT, OH MY!

Last year on vacation in Athens, Greece, I decided purchase a bust of the great Greek geometer Pythagoras, thinking by osmosis I might be the beneficiary of a few residual crumbs of ancient Grecian learning. All day long I rummaged through the tiny shops of the *Plaka*, looking amid thousands of busts and statues of every Greek god and goddess since Zeus and Artemis.

Finally, after a day of no avail, a shopkeeper said in an enlightened tone, "Oh, you mean *Pith-a-GOR-us.*" This unusual pronunciation of one of the famous names in geometry seemed rather odd since I have been saying "puh-THA-guh-rus" for the past century and a half in my geometry classes. But whatever the reason for his weird pronunciation, he led me to a back room filled with hundreds of busts of the Great Geometer of Samos. Although I didn't think anything about the incident at the time, after returning home I started thinking about names of famous mathematicians, or more accurately, their correct pronunciation.

If I had a dime for every time one of my students mispronounced the name of a famous mathematician, I would have turned in my chalk and eraser ages ago.

When I lecture about the works of the great Swiss mathematician of fame, I repeat again and again, "The name is OY-ler." OY-ler, which translates at the end of the semester to a student asking if YEW-ler's theorem will be on the final exam.

Maybe I should have a more relaxed disposition when it comes to student's creative renderings of the names of the greats since as a student, I too, mangled my share of surnames.

But all that was before I took Mr. Kronecker's high-school calculus class. 'Ol Krony was a good enough teacher, but he had one major flaw. He was a drill sergeant when it came to the history of mathematics and its famous contributors, which naturally implied, pronouncing them correctly.

No, no, no, Krony would scream at some poor soul as he or she butchered the name of the Frenchman responsible for l'Hospital's Rule. Noooooooooooo, he'd yell as the poor student continued coughing up 'Las-Hoopie'. 'Lew Hoopertal', 'Lo-Hippie', ...'La-HOSPITAL!'?

By this time Krony would be so angry he'd get out his key chain and begin whirling it around. Students in the front row would look on anxiously, knowing they might be at risk from a flying car key. The hapless student in Krony's crosshairs probably didn't learn the importance of l'Hospital's Rule, but he positively learned to say it right.

Another time 'ol Krony's key chain got a work-out was when I gave a talk on the contributions of the eminent French mathematician, Lagrange. My grandparents were from Lagränge , Illinois so the question of pronouncing the guy's name never entered my mind. I barely finished the title of my talk when I heard the familiar drone of Krony's key chain. By the time I got to the next Lagrange, I was shaking so hard I could only say, `La, La, uh, er, that Spaniard that invented the derivative. That was the last time anyone ever saw Krony's keys. They broke clean off and went right out the window. He never did find them.

Another incident that took years off Krony's life was when he tried to teach Truck Turley the correct pronunciation of the great Swiss mathematician, Leonhard Euler. Truck was our school's star football player, but never applied himself to mathematics.

"*AY-ler, YOO-ler, YEW-ler,*" Truck was running

through various combinations. Krony was determined Truck should say it correctly since the big homecoming game was coming up and Krony didn't want anyone to think he was passing out favors to athletes.

"EYE-lay," the big fullback grunted. By now Krony was searching his pockets for his favorite key chain. However, he wasn't about to let Truck miss this one. Finally, after several minutes of listening to all the O's and E's in the dictionary, Krony leaned over Truck and whispered ...

"OINK,OINK,OINK."

The class leaned forward.

"OY'-LER!" Truck burst out.

"That's right!" Krony said as he let out a sigh of relief, "and now for the chain rule."

Krony never recovered from the "Oink, Oink" incident. Principal Whipple, heard about the story through the grapevine and gave Krony a reprimand.

Krony once gave our class a handout giving the correct pronunciation of many of the great mathematicians throughout history that are often mispronounced.

————¤¤¤☰☰☰¤¤¤————

Pronunciation of Names of Mathematics

1. Abel, Niels (AH-buhl)
2. Agnesi, Maria Gaetana (an-YAY-zee)
3. al-Khwarizmi (al-KWA-riz-me)
4. Archimedes (ahr-ki-MEE-deez)
5. Banach, Stefan (BAH-nahkh)
6. Bernoulli, Johann (bur-NOO-lee)
7. Bernoulli, Jacob (bur-NOO-lee)
8. Bolyai, Janos (BAHL-yoy)
9. Bolzano, Bernard (bohlt-SAH-noh)
10. Cardano, Gerolamo (kahr-DAH-no)
11. Cauchy, Augustin-Louis (KOH-shee)
12. Clairaut, Alexis (kluh-ROH)
13. d'Alembert, Jean (DAH-lahm-behr)
14. Decartes, Rene (day-CART)
15. Dedekind, Richard (DAY-du-kint)
16. Desargues, Gerard (day-ZARG)
17. Dirichlet, Lejeune (dee-ree-SHLAY)

18. Euclid of Alexandria (YOO-klid)
19. Euler, Leonhard (OY-ler)
20. Eratostheses (er-a-TOS-then-eez
21. Fermat, Pierre de (fehr-MAH)
22. Fibonacci (fe-bo-NAHT-chee)
23. Fourier, Joseph (foor-YAY)
24. Frobenius, Ferdinand (fro-BEE-nee-us)
25. Frege,Gottlob (FRAY-guh)
26. Galois, Evasiste (gal-WAH)
27. Gauss, Carl Friedrich (GOWSS)
28. Gödel, Kurt (GUR-duhl)
29. Hermite, Charles (ehr-MEET)
30. l'Hospital, Marquis de (loh-pee-tal)
31. Huygens, Christiaan (HOY-guhnz)
32. Jacobi, Carl (juh-KOH-bee)
33. Kronecker, Leopold (KRON-nek-uhr)
34. Lagrange, Joseph-Louis (luh-GRAHNZ)
35. Laplace, Pierre-Simon (lah-PLAHSS)
36. Lebesgue, Henri (luh-BAG)
37. Leibniz, Gottfried Wilhelm (LYB-nits)
38. Liouville, Joseph (LYOO-veel)
39. Lobachevski, Nikolai (loh-buh-CHEF-skee)
40. Mobius, August Ferdinand MOW-bee-us]
41. Monge, Gaspard (MAHNZH)

42. Noether, Emile (NOH-eh-thir)
43. Poincare, Henri (pwahn-kah-RAY)
44. Pythagoras (puh-THA-guh-rus)
45. Ramanujan, Srinivasa (raw-maw-nu-juhn)
46. Ricatti, Jacopo (ree-KAH-tee)
47. Riemann, Bernhard (REE-mahn)
48. Tartaglia (tahr-TAG-lia)
49. Tchebycheff, Pafnuti (CHEB-ih-sheff)
50. Thales of Miletus (tha-LEES of Mi-LEE-tus)
51. Torricelli, Evangelista (tohr-ree-CHEL-lee)
52. von Neumann, John (fon NOY-mahn)
53. Weierstrass, Karl (VY-uhr-shtrass)
54. Wiener, Norbert (VEE-nuhr)

$$\Pi\psi\Delta\Phi\theta\Lambda$$

18

WHAT IS MATH ? STRAIGHT FROM THE HORSES' MOUTH

There are so many myths about mathematics you'd think it was a religious cult — and the biggest myth of all is that there are two types of people, "math people" and "non-math people."

For many non-math people" being ignorant of math is taken as a badge of honor. "I'm so bad in math, I can't even make change," is a comment you often hear someone announce proudly. So now when a non-math person asks me what I do for a living, I tell them I'm trying to find x." If they tell me that they teach English, I tell them I'm so bad at reading, if it weren't for those little pictures on the doors, I couldn't tell the men's room from women's room.

Sometimes a non-math person ask me if there are applications of mathematics. I tell them the story of when I was working on my Ph.D thesis and trying to find applications of the Schauder fixed point theorem. I had just met my future wife who knew nothing about mathematics, but I managed to explain the famous fixed-point theorem to her in a language she could understand. She always nodded in agreement with everything I said so I was impressed with her understanding. Later, after we were married she told me she didn't know what the hell I was talking about, but was so impressed with my mathematical diligence that she continued to date me. So you see, sometimes you can find applications of mathematics in the least likely places.

—¤¤¤ΞΞ¤¤¤—

So, why is it so many people are turned off to mathematics? My own theory is it's the "piano syndrome." Learning to play the piano and learning elementary mathematics have a lot in common. In both endeavors, the beginner must pass through a rigid orientation. The beginning pianist spends months developing finger dexterity by playing scale after scale. In mathematics the child must first learn to count, then arithmetic, algebra, and so on. In both cases, such activities are considered by most people to be dull and monotonous. Only after the basics are mastered can the student of the piano interpret a Chopin concerto and the student of mathematics use one's imagination to build mathematical worlds. It is unfortunate that beginning students of mathematics never survive their rigid introduction to experience the exciting and wonderful mathematics world that lies beyond.

What better way to throw off your old misconceptions of mathematics and learn its true meaning straight from the horse's mouth — by perusing the thoughts and opinions of the greatest mathematicians who ever practiced the Queen of the Sciences.

If people do not believe that mathematics is simple, it is only because they do not realize how complicated life is. — *John von Neumann*

Mathematics is one of the essential emanations of the human spirit – something to be valued in and of itself, like art or poetry. — *Oswald Veblen*

Life is good for only two things, discovering mathematics and teaching mathematics. — *Simeon Poisson*

A mathematician who is not also something of a poet will never be a perfect mathematician. —*Karl Weierstrass*

The mathematical sciences particularly exhibit order, symmetry, and limitation and these are the greatest forms of the beautiful. — *Aristotle*

Nature is written in mathematical language. —*Galileo Galilei*

Mathematics is the science of what is clear by itself. — *Carl Gustav Jacob Jacobi*

It is easier to square the circle than to get around a mathematician. *Augustus De Morgan*

We are servants rather than masters in mathematics. — *Charles Hermite*

No matter how correct a mathematical theorem may appear to be, one ought never to be satisfied that there was not something imperfect about it until it also gives the impression of being beautiful. — *George Boole*

In mathematics the art of proposing a

question must be held of higher value than solving it. — *Georg Cantor*

Every good mathematician is at least half a philosopher, and every good philosopher is at least half a mathematician. — *Gottlob Frege*

It is impossible to be a mathematician without being a poet in soul. — *Zofia Kowalewska*

Mathematicians are born, not made. — *Jules Henri Poincaré*

Mathematics knows no races or geographic boundaries; for mathematics, the cultural world is one country. — *David Hilbert*

Mathematics is the most beautiful and most powerful creation of the human spirit. — *Stefan Banach*

Perhaps the most surprising thing about mathematics is that it is so surprising. The rules which we make up at the beginning seem ordinary and inevitable, but impossible to foresee their consequences. They have only been found by long study extending over many centuries. — *E. C. Titchmarch*

A mathematician is a machine for turning coffee into theorems. — *Paul Erdös*

The only way to learn mathematics is to do mathematics. — *Paul Halmos*

In mathematics you don't understand things you just get used to them. — *John von Neumann*

The definition of a good mathematical problem is the mathematics it generates rather than the problem itself. — *Andrew Wiles*

Everyone knows what a curve is until he has studied enough mathematics to become confused through the countless number of possible exceptions. — *Felix Klein*

Let us grant that the pursuit of mathematics is a divine madness of the human spirit, a refuge from the goading urgency of contingent happenings.—*Alfred North Whitehead*

The moving power of mathematical invention is not reasoning but imagination.— *Augustus de Morgan*

As in mathematics, so in natural

philosophy, the investigation of difficult things by the method of composition ought to proceed the method of composition. — *Sir Isaac Newton*

Algebra is the offer made by the devil to the mathematician. All you need to do, is give me your soul: give up geometry. — *Michael Atiyah*

Mathematics is the music of reason. — *Henri Poincare'*

In my opinion all things in nature occur mathematically. — *Rene Decartes*

There is no study in the world which brings into more harmonious action all the faculties of the mind than mathematics. — *James Sylvester*

It is a platitude that pure mathematics can have unexpected consequences that affect our daily lives. — *J. E. Littlewood*

ΣΞΔΘΩΨ

19

BASEBALL'S MAGIC NUMBER: SMALLER IS BETTER

I'm not one to bloviate and prattle on about myself but back in my day I was what sports-writers referred to as a sports phenom. I had the agility of a gymnast, the strength of a weight lifter, the endurance of a marathon runner, and the imagination of J.K Rowling. Especially the imagination, I even imagined I made the team once in a while.

But on the other hand, they did retire my athletic number, but in my case they retired it *before* I got to high school. You see, there was a slight erratum in my poop sheet. I wasn't a sport's nobility, I was a sport's *nobody*. I wasn't an athletic force, I was an athletic *farce*. I wasn't an idol on the field, I was *idle* on the field. I didn't soar on the field, I was *sore* on the field. Ok, ok you get the idea. I was `ol butterfingers himself, mister stumblebum, `ol lead-foot.

But although my physical attributes never made all-conference, I was able to overcome my athletic failings in a way that would make the average bench-warmer proud. I learned more baseball statistics than an entire 9-man team, a fact the average 9-man team came to regret. My position as batboy offered me the ideal spot to annoy player after player with an endless stream of trivial minutiae. "Do you know the National League player with the most prime number batting averages?" I'd ask the league-leading home-run hitter as he took his warm-up swings in the on-deck circle. Or maybe, "Did you know Boston's magic number against the Yankees is down to 113?"

I generally rattled off enough baseball trivia that by the time the batter got in the batter's box, the poor guy couldn't hit a beach ball with a tree trunk.

Now, I can understand baseball fans not knowing the National League player with the most prime number batting averages, but not knowing the magic number of your favorite team at any time during the season? Tsk tsk.

So, what is baseball's magic number and how is it computed?

Baseball's magic number of one team, say Team A, against another team, say Team B, at any point in time during the season, is any combination of additional wins for Team A, or losses by Team B, after which Team A is ensured of finishing ahead of Team B at the end of the season.

For example, consider teams, Team *A* and Team *B* illustrated in the following box, where Team *A* leads Team *B* by 4 games in the win column, and 5 games in the lost column (ahead by 4 and a half games in baseball lingo). Assume the teams are major league teams, where each team plays a 162-game schedule, which means Team *A* has 7 games left to play and Team *B* has 6 remaining games.

If you analyze the situation carefully, you will see that Team *A*'s magic number against Team *B*'s is 3, meaning any combination of 3 wins for Team *A* or losses for Team *B* will ensure Team *A* of being ahead of Team *B* when the season ends. Note that even if Team *B* wins all 6 remaining games, Team *A* will still come out on top if it wins at least 3 of its remaining 7 games. (Team *A* will end the season at 91-71, whereas Team *B* ends at 90-72.)

	W	L	Games Left
Team A	88	67	$162 - 88 - 67 = 7$
Team B	84	72	$162 - 84 - 72 = 6$

148

Although a lower ranked team's magic number against a higher ranked team is not normally quoted, it is a worthwhile statistic and just as easy to compute. For the above teams, Team B's magic number against Team A is 12, meaning any combination of 12 Team B wins and Team A losses will ensure Team B comes out ahead (not just tied) of Team A at the end of the season. The reader can try some different end-of-season scenarios to verify 12 is the correct number.

The Formula for Baseball's Magic Number

To find the magic formula for the magic number of Team A against Team B, simply observe the basic fact that Team A will finish ahead of Team B at the end of the season if at the end of the season it wins more games! It is a simple as that. To put this fact into finding the formula, assume at some point during the season Team A has the record of $W_A : L_A$ and Team B has a record of $W_B : L_B$ where W's represent wins and L's losses of the two teams at that point during the season. We also denote the number of games in the sea-

son by G, where for Major League baseball it is
162.

	W	L	Games Left
Team A	W_A	L_A	$G - W_A - L_A$
Team B	W_B	L_B	$G - W_B - L_B$

Suppose now for the *remainder* of the season
Team A has a won:lost record of w_A, l_A and
Team B has a won:lost record of w_B, l_B . Of
course, we don't know these values at any time
during the season.

To find Team A's magic number against Team B,
we use the simple observation that Team A will
finish ahead of Team B if it wins one more game.
Stated mathematically, that says

$$(W_A + w_A) - (W_B + w_B) = 1$$

(Of course Team A could win *many* more games
than Team B, but winning just one more game is
sufficient for Team A to come out on top.)

But $W_B + w_B = G - L_B - l_B$ (right?) and so plugging this into the previous equation, we have

$$(W_A + w_A) - (G - L_B - l_B) = 1$$

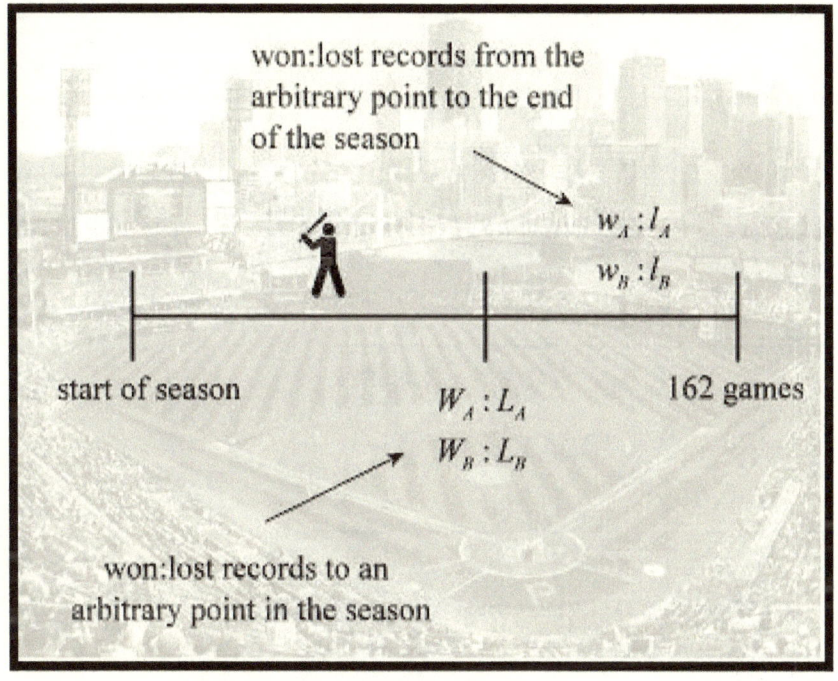

won:lost records from the
arbitrary point to the end
of the season

$w_A : l_A$

$w_B : l_B$

start of season

$W_A : L_A$

162 games

$W_B : L_B$

won:lost records to an
arbitrary point in the season

Baseball is a statistician's motherlode

so solving this simple equation for $w_A + l_B$, which we do not know at any time during the season, but solving it we find a number which *can* be computed at any time during the season and this number is baseball's magic number for a 162 game season:

$$\text{Magic Number: Team } A \text{ vs Team } B = w_A + l_B$$
$$= G + 1 - W_A - L_B$$
$$= 163 - W_A - L_B$$

For example consider the following records of Team A and Team B in the middle of a 162 game season,

	Wins	Losses
Team A	53	33
Team B	43	40

The magic numbers of each team versus the other are as follows:

- Team A vs Team B = $163 - 53 - 40 = 70$
- Team B vs Team A = $163 - 43 - 33 = 87$

———¤¤¤☷☷☷¤¤¤———

For example, suppose on August 25 the stand-ings in the Eastern Division of the American League are as follows:

	W	L	Games Left
Boston	69	41	52
New York	53	56	53
Baltimore	51	59	52
Toronto	47	61	54
Detroit	46	63	53

At this point in time, Boston's magic number against each of the trailing teams is

- Magic # vs New York = 163-69-56 = 38

- Magic # vs Baltimore = 163-69-59 = 35

- Magic # vs Toronto = 163-69-61 = 33

- Magic # vs Detroit = 163-69-63 = 31

My-Wins-Your-Losses Number: For you baseball fans let me give you a tip. Forget baseball's magic number. It's simpler to compute the

my-wins-plus-your-losses

number. Here's the idea. Suppose you are a Red Sox fan and have a lifetime hatred of the Yankees. That means you are happy whenever the Sox win and the Yankees lose. Hence the statistic you should be constantly computing is the number of the Sox wins plus Yankee losses.

At the start of the season this number is obviously zero (no Sox wins, no Yankee losses), but when (hopefully) this number reaches 163, then you can be sure the Sox will finish ahead of the Yankees at the end of the season. (It's really just the same as saying the magic number reaches zero.). For example suppose on Sept 1 the Sox and Yankees have the following records.

	W	L
Yankees	75	50
Red Sox	72	53

This means the Sox-wins-plus-Yankee-losses number is 72 + 50 = 122, which means that another combination of 41 Sox wins and Yankees losses will assure the Sox will be ahead of the Yankees at season's end.

ΠψΔΦθΛ

20

A 49-PROVERB ODYSSEY THROUGH MATHEMATICS

———

Someone once famously said

"Why are numbers beautiful? It is the same as asking, why is Beethoven's Ninth Symphony beautiful? If you don't see why this is so, no one can tell you."

In fact, the quote was the passionate belief of the Hungarian mathematician Paul Erdõs, who was famous for his introspections of mathematics.

Another observation by one of the great minds of the Queen of Science was due to the Russian mathematician Nikolai Ivanovich Lobachevsky, who observed

> *"There is no branch of mathematics, however abstract, which may not some day be applied to phenomena of the real world."*

It can be argued that mathematics has allowed us to understand the world more than any other intellectual discipline.

Reading people's thoughts about mathematics provides a glimpse into the nature of the beast. Whether you agree or disagree, we pass off a few viewpoints from the massive pile of opinions that have been floating around the mathematical atmosphere for over 2500 years, possibly even before Thales of Miletus realized the importance of the mathematical proof.

—¤¤¤Ξ Ξ¤¤¤—

1. Mathematics is the last discipline that comes to a developing country. It is only after everything else is developed that people turn to abstract thought. Say what you will about mathematics, at least it is right.

2. Mathematics is the most democratic of all disciplines. No one can discriminate against the laws of logic.

3. Mathematics is like an oak tree. Great things come from small beginnings.

4. Religious differences breed wars, mathematical differences breed new ideas.

5. After all is said and done, mathematicians are deduction robots.

6. Mathematics and science are the yin and yang of knowledge. Mathematics is deduction and science is induction.

7. Mathematics is the keeper of infinity.

8. A good theorem is like sausage and laws. The finished product may be a delight, but you wouldn't want to see it being made.

9. In the long run, one can better defend a country by teaching its children mathematics than by building tanks.

10. Mathematics waters the imagination as spring rains water a budding flower.

11. In mathematics you don't understand things, you just get used to them.

12. Eventually mathematics will become a necessity and then we mathematicians will rule the world.

13. The brightest ideas of mathematics are just that until they are proven.

14 The test of an axiom lies in the theorems it produces.

15. A mathematician must believe in both the possibility of God and the Devil.

16. In deciding whether to become an engineer or a mathematician, one must decide whether one likes to do or to think.

17. To learn mathematics is to acquaint oneself with the best that mankind has to offer.

18. In mathematics, only the genius gets lucky.

19. The whole of mathematics is not comprehensible to any one person.

20. To anyone not instructed in mathematics, the world must seem like a wonderful place, much of which is hidden from view.

21. I would rather have a dozen root canals than to work a single word problem.

22. Say what you like, there is no demagoguery in mathematics.

23. A good mathematician is one who makes the smallest amount of ideas go a long way.

24. Mathematics is the art of drawing necessary conclusions from sufficient premises.

25. The work of a great mathematician will effect future generations, but even he knows not where his theorems will lead.

26. The problem that makes the study of integers so hard is that they are so simple. What we must do is make them complicated.

27. Mathematics, it's just one damn theorem after another.

28. A person who is both a mathematician and a poet has it all.

29. The difference between a mathematician and a politician is that a mathematician tries to say the most with the least number of words, a politician does the opposite.

30. If mathematics were dogmatic, there would be no mathematics.

31. Mathematicians make their own language so no one else can understand them.

32. The old is never destroyed in mathematics. It is the foundation for new ideas.

33. An axiom is intuition that has passed the test of time.

34. If mathematicians would only realize how much they bore everyone.

35. One way to end a conversation is to tell everyone you're a mathematician.

36. Mathematics is not interested in race, creed, or religion. It is interested only in mathematical truths.

37. Now really, just where would we be without the Greeks?

38. Mathematics is the highest rung of human thought.

39. In all reality mathematics is beyond our mental facilities to understand it.

40. How can a finite mind comprehend the infinity of mathematics?

41. Mathematicians are the soothsayers and witch doctors of the 21th century.

42. Mathematical history is in essence a history of great ideas.

43. You can tell where a nation is going by the mathematicians it produces.

44. Mathematics is simply ideas reduced to their ultimate essence.

45. Obvious is the most dangerous word in mathematics.

46. Mathematics from the right vantage point possesses not only truth but beauty.

47. Mathematics is music for the mind.

48. Mathematics may just well be the language of God.

49. Mathematics may just well be the language of God.

ΠΦΓΛΘΩ

21

A PHYSICS PROBLEM SOLVES A MATH PROBLEM

———

A Physics Problem Solves a Math Problem

I wandered lonely as a cloud,
That floats on high o'er vales and hills.
When all at once I saw a crowd,
A host of golden daffodils.
... William Wordsworth

You see what's going on here, don't you? In the poem Daffodils, Mr. Wordsworth is celebrating the wonders of nature through the medium of rhymed verse. Musicians, artists, photographers as well as poets have all applied their genius to display the beauty of the real world through their respective mediums: Vivaldi's *Four Seasons*, Hokusai's *The Wave*, as well as Ansel Adams' photographs of the American West allow us to experience the world through new eyes.

But as much as our spirits are lifted by artistic expression, when it comes to untangling the riddles of the universe, nothing comes as close to the task as mathematics.

The English writer G. J. Chesterton once observed

> *"The difference between a poet and a mathematician is that the poet tries to get his head into the heavens while the mathematician tries to get heaven into his head."*

Although one cannot help but be stirred by the stark imagery in Hokusai's famous painting *The Wave* which depicts an enormous threatening wave, no one would argue that the motion of waves is more accurately described by the Navier Stokes equations which are equations that govern the motion of fluids.

$$\frac{\partial \eta}{\partial t} + \frac{\partial (\eta u)}{\partial x} + \frac{\partial (\eta v)}{\partial y} = 0$$

$$\frac{\partial \eta}{\partial t} + \frac{\partial}{\partial x}\left(\eta u^2 + \tfrac{1}{2}g\eta^2\right) + \frac{\partial (\eta v)}{\partial y} = 0$$

$$\frac{\partial \eta}{\partial t} + \frac{\partial}{\partial x}\left(\eta u v\right) + \frac{\partial}{\partial y}\left(\eta v^2 + \tfrac{1}{2}g\eta^2\right) = 0$$

The great scientist Albert Einstein once asked how mathematics, a product solely of human thought, can so accurately unveil the mysteries of nature. Mathematical modeling, or the description of objects of reality through equations, has long been a staple of scientists of all stripes, from physicists, and chemists, to biologists and even social scientists.

But one wonders if the tables can be turned? Can pure mathematical problems be solved with the assistance of objects of reality? The answer is yes and one need look no further than analog computers, which are made up of circuitry that simulates and hence solves (or approximates the solutions of) differential equations.

A fascinating physical apparatus that is able to solve algebraic equations was described by Italian mathematician I. Ghersi in his famous book *Matematica dilettevole e curiosa*. The apparatus, shown in Figure 1, consists of cone and cylinder containers that are connected by a small pipe which allows water to flow from one container to the other.

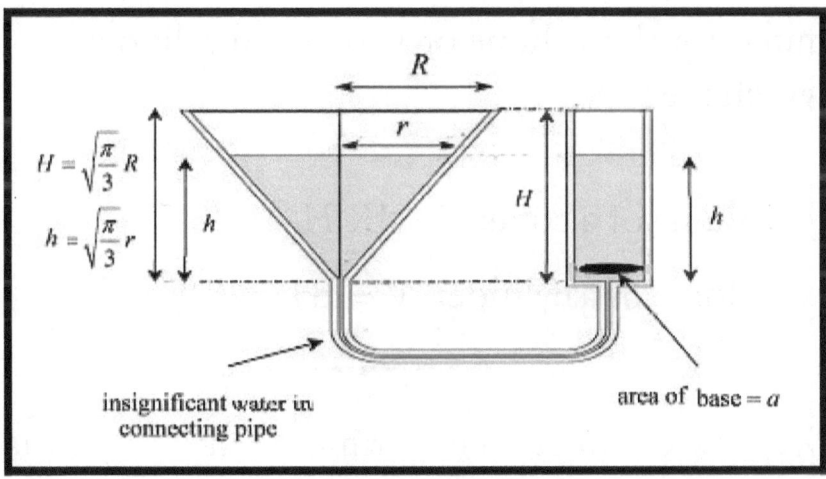

$$H = \sqrt{\frac{\pi}{3}} R$$

$$h = \sqrt{\frac{\pi}{3}} r$$

insignificant water in connecting pipe

area of base $= a$

Figure 1: Cone – Cylinder equation solver

We denote the height of the cone by H and its

radius by R. The cone can be any size but the height is slightly larger than the radius, the exact relation between them being

$$H = \sqrt{\pi/3}\ R$$

On the other hand the cylinder is a normal cylinder with a circular base, where we denote the area of the base by the letter a.

To understand how this cone-cylinder apparatus is able to find a solution of an algebraic equation, it is necessary that we know the formulas for the volume of a cone and cylinder, which are

---

Volume of a cone: $V = \pi R^2 H$

Volume of a cylinder: $V = aH$

We now conduct an experiment whose result is the solution of a particular equation. We begin by pouring c cubic inches of water into either the cone or cylinder. Observe that the water level h in the cone or cylinder will be the same.

You can choose the volume of water added in any amount desired as long as you don't overflow the apparatus.

We now measure the amount of water in each container, recalling that the height/radius restriction of the cone satisfying

$$H = \sqrt{\tfrac{\pi}{3}} R \Rightarrow R^2 = \left(\tfrac{3}{\pi}\right) H \Rightarrow r^2 = \left(\tfrac{3}{\pi}\right) h$$

resulting in the following values

Volume of water in a cone: $V_{cone} = \pi r^2 h / 3$

Volume of water in a cylinder: $V_{cylinder} = ah$

But the amount of water added to the apparatus is c cubic inches and so adding the water from the cone and apparatus yields the equation

$$h^3 + ah = c$$

In other words we can find a solution h of this cubic equation by measuring the height h of the water in either the cone or cylinder. We can do this for different values of the constant c (the cubic inches of water added) and constants a (the area of the base of the cylinder.

Suppose we wish to find a solution to the cubic equation

$$h^3 + h = 15$$

We construct a cylinder with base area $a = 1$ cubic inches and add $c = 15$ cubic inches of water to either the cone or cylinder. The water level h satisfies the equation. In this case about $h = 2.3$ inches.

———¤¤¤ΞΞΞ¤¤¤———

Newer Models

Readers familiar with calculus will realize our apparatus in Figure 1 is only Model 1 in a long line of possible *apparati* that could be constructed to solve all sorts of algebraic equations. For example if we replace the cone by a container whose sides consist in rotating a curve defined by the function $y = f(x)$ around the y-axis as shown in Figure 2, we find the volume of water in the container (using the method of discs, remember?) rising to a height h of

$$V_{object} = \pi \int_0^h f^2(y)\, dy$$

The apparatus is demonstrated in the following diagram.

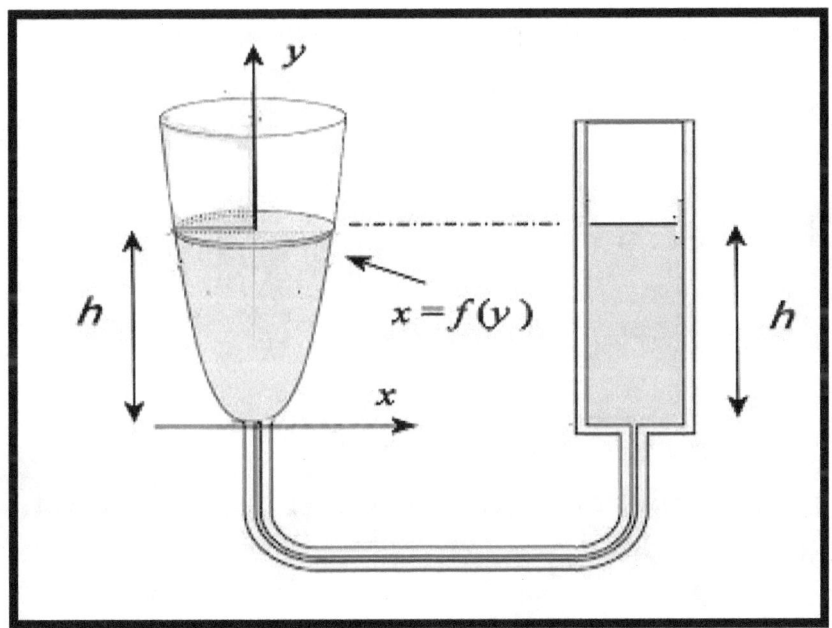

Figure 2: Mechanical device for doing calculus

Pouring c cubic inches of water into either the

new container or cylinder, the common water height h will satisfy the equation:

$$\pi \int_0^h f^2(y) \, dy + h = c$$

both as a term in the equation and as the upper limit of a definite integral.

If we select the side of the container as the curve $y = x^2$ or equivalently

$$x = f(y) = y^{1/3}$$

then the container contains

$$V_{container} = \pi \int_0^h f^2(y) \, dy = \pi \int_0^h y^{2/3} \, dy = \tfrac{3\pi}{5} h^{5/3}$$

cubic inches and combining this with the ah cubic inches in the cylinder, the total amount of water in the apparatus is

$$\tfrac{3\pi}{5} h^{5/3} + ah = c$$

cubic inches. Hence, by measuring the height h in this apparatus after c cubic inches of water are added, we have found a solution to the above equation, one which contains a *fractional expo-nent*, no minor feat.

$$\Pi\psi\Delta\Phi\theta\Lambda$$

22

YES VIRGINIA, THERE IS LIFE AFTER MATHEMATICS

I was taken back a while ago when a student came in my office and instead of asking the standard questions of what expect on a forthcoming exam or requesting to turn in a late homework, asked a rather curious question.

"Do you remember an old student of yours from about 25 years ago?" she said.

Normally, I don't remember the names of students I had the previous semester, but for some reason I remembered this one student.

"Why yes, I remember her," I admitted.

"*She's my mom!*" she blurted out.

And to make matters worse, the incident didn't take place last week. *It was twenty years ago!* So to make sure another student doesn't come rushing in my office and ask me if I remember so-and-so, then blurt out, "She's my *grandma!*" I decided to take the safe route, retire, and, as they say, get the hell outta Dodge.

Conventional wisdom about mathematicians is they're virtuosos at mathy things, but utterly worthless at everything else, which as general principles go, is a general principle. I know of mathematicians who can't screw in a light bulb, and others who are damn good electricians, carpenters, and plumbers. I know the latter for a fact, since it's them I generally call.

Recently, I decided to get a life outside of mathematics and so I became a volunteer with a local ambulance rescue squad. It was really quite simple. I took a boy-scout course in life-saving, learned how to wrap a leg in a splint, drive an ambulance, and off I went. It wasn't a medical-school course at Johns Hopkins.

I could tell you dozens of stories about my life in the rescue business, but I'd like to tell you about a particular run we had recently on Christmas Eve.

I always volunteer for duty on Christmas Eve. It seems something interesting always comes up.

It was 2:30 A.M. Christmas morning, and I had just settled down for a long winter's nap when all of a sudden I stirred like a flash. No silly, it wasn't a miniature sleigh and eight tiny reindeer, my beeper went off. But I did spring from my bed and pull on the first thing I could find. Arriving at the ambulance bay, we learned an old woman had fallen and possibly broken a shoulder.

Arriving at the old woman's house, things were unusual to say the least. A police car, which arrived at the house ahead of us, was rammed by a hit-and-run driver, which we later learned, was a drunk, mad at the policeman for breaking up a domestic dispute earlier in the evening. Like I said, it was Christmas Eve.

Now, friends, we are in the state of Maine and the temperature was -25°, which meant we were attired for the worst. I had grabbed a couple sweaters, pants, overcoat and my wife's red stocking cap. My two colleagues also seemed to have dressed in haste as one of them was wearing a pair of pajamas sticking out under an old overcoat, the other had a pair of Long-Johns tucked inside a pair of overshoes.

By the time we unloaded our gear and rushed into the house, things were starting to get even more strange. The policeman, whose car was rammed, was injured but managed to make it to the house. He was lying on the sofa in front of a Christmas tree with several people gathered around him.

So now we had two emergencies. Sensing the policeman was the more serious, my more experienced colleagues rushed to his aid and told me to attend to the old woman.

So, off I go, bag in hand, to find a very old woman, sitting alone at the side of her bed. I thought she couldn't be hurt that much since she looked at me in my stocking cap and asked if I was one of the Three Wise Men. She said she saw a strange threesome through her window running up the driveway. I told her I was a medical professional, and my bag didn't contain gold, frankincense and myrrh. I told her to just relax while I took her blood pressure and pulse, just as my training prepared me. She didn't bother to tell me she had been an emergency nurse for forty years.

Just as I was putting the old woman's arm in a sling, I heard commotion coming from downstairs, telling me to forget the old woman and get down there pronto. It seemed the policeman was going into medical shock. He had a river of sweat running down his face and was shaking like a leaf. His only request was to be shoved into an oven at 350 degrees. He had no such luck however. We strapped him to the gurney and rushed him outside. Unfortunately, the driveway was covered with a sheet of ice and we all started to slide down the long driveway.

I was hanging onto the front of the gurney while my two colleagues were hanging onto the back. We got halfway down the driveway and the gurney turned, sideways at first, and then all the way around. My colleagues were now at the front and I'm hanging onto the back. I'm thinking we're going to crash into the ambulance, killing us all. However, just before we got there we made another 180° turn and stopped right at the back doors of the ambulance.

It was perfect. We just opened the bay doors, set in the policeman, and rushed off to the hospital. The policeman was a real sport about the whole incident. He never said a word.

The next morning, the officer was at home with

his wife and kids, the only lingering side effects being minor psychological ones from the trip down the driveway.

Meanwhile, someone called for a back-up ambulance for the old woman. I understand she watched the whole episode of the "big slide" from her bedroom window. She told the back-up crew it looked like Santa Claus and three motley reindeer.

I understand too she was the one who requested the backup.

$$\Pi\psi\Delta\Phi\theta\Lambda$$

www.ingramcontent.com/pod-product-compliance
Lightning Source LLC
Chambersburg PA
CBHW021425170526
45164CB00001B/103